Mathematics:
Applications and Interpretation HL
FOR THE IB DIPLOMA

Ian Lucas

PEAK
STUDY RESOURCES

Published by:
Peak Study Resources Ltd
1 & 3 Kings Meadow
Oxford OX2 0DP
UK

www.peakib.com

Mathematics: Applications and Interpretation HL
Study & Revision Guide for the IB Diploma

ISBN 978-1-913433-03-1

Peak Study & Revision Guides for the IB Diploma have been developed independently of the International Baccalaureate Organization (IBO). 'International Baccalaureate' and 'IB' are registered trademarks of the IBO.

Orders: books may be ordered directly through the publisher's website www.peakib.com, or to enquire about local stockists please contact us at books@peakib.com (schools@peakib.com for educational establishments).

Printed and bound in the UK by CPI Group (UK) Ltd, Croydon CR0 4YY

www.cpibooks.co.uk

Contents

About this book

This is a study guide, not a text book. Its aim is to review each part of the subject and ultimately prepare you for your exams. Covering the complete syllabus, I show you how the topics can translate into questions, giving you plenty of tips and shortcuts.

The exam is not so much a direct test of your knowledge and understanding (you will not get a question which begins "What do you know about...?"); but a test of how you use your knowledge and understanding to solve mathematical problems. So the emphasis in this book is on how to answer questions. In particular you will find plenty of fully worked examples in the text (those similar to Paper 1 exam questions are shaded green), as well as further exercises where it is useful to see how questions can probe a topic from different angles. At the end of each chapter you will find longer questions which are similar in style to those in Paper 2 in the exam – by their nature, such questions may need knowledge of several different areas of the syllabus – and these are indicated with a red band. Answers are provided in the book, but for full working you will need to visit the Mathematics area of the Peak Study Resources website at www.peakib.com.

You are expected to be able to understand and use your graphic display calculator (GDC) to the full in many areas of the syllabus. Indeed, some questions require you to use, for example, the graphing or equation solving features. Since different people use different calculators, it is not possible for this book to explain the detail of their use; but I have indicated calculator tips (using the symbol ▦), and also questions which require calculator use. The more you can use your GDC, the more proficient you will become.

This book is just one resource that will help you prepare for your exams; another is the set of short videos on the website which lead you through the working and solutions of a range of exam-style questions. This area of the website is updated with new content as we add more videos to the resource bank. I have indicated in the text if a question has a video solution: look for the video symbol ▶.

I have liberally splashed notes boxes in the margins throughout the book. These contain hints, warnings, exam tips, "dos and don'ts", suggestions... do read them all, as well as the blue text in the question boxes. There's so much information which can help you with those precious extra marks. And there are similar yellow boxes which contain links to other pages, websites, videos blogs.

Question (Paper 1 style)	
Worked answer	*Hints and tips*

Question (Paper 2 style)
———
Answer

Notes box – please read what they contain!

Reference box; leads you to useful resources.

Acknowledgements

I am enormously grateful to Peter Gray, an IB examiner currently at Munich International School, who has proof read this book and, in the process, made some eminently sensible suggestions for numerous improvements; he has also tactfully pointed to a number of errors in both the text and the calculations which I have gratefully corrected! Any remaining errors are entirely my responsibility, and I would be very happy to hear from readers who find any further errors, or who have suggestions for improvements.

Through teaching with Oxford Study Courses over the last 20 years, I have been privileged to help many students revise for their IB Mathematics exams, and much of what I have learnt has been distilled into this book. I would value any feedback so that later editions can continue to help students around the world. Please feel free to e-mail feedback@peakib.com, or comment via the normal social media channels.

Ian Lucas

Chapter 1: NUMBER AND ALGEBRA

1.1 Number Systems

Different situations require different types of number. For example, populations of countries will always be given as positive whole numbers, whereas the division of a reward will require the use of fractions. These are known as *number systems*, and the ones you need to know are:

- *Natural numbers* – positive whole numbers.
- *Integers* – whole numbers including negatives and zero.
- *Rationals* – numbers which can be written as fractions.
- *Irrationals* – numbers which can't be written as fractions.
- *Reals* – the rationals and the irrationals put together. The reals will include every possible number you could meet in the course.

The diagram below shows how the sets are related to each other. For example, every integer can be written as a rational (such as $4 = \frac{4}{1}$), so integers are a subset of rationals.

> You also need to know the conventional symbols used for the main number systems:
>
> $\mathbb{N}$ = Natural numbers
> $\mathbb{Z}$ = Integers
> $\mathbb{Q}$ = Rational numbers
> $\mathbb{R}$ = Real numbers
> $\mathbb{Z}^+$ = Positive integers

Reals = Rationals ∪ Irrationals

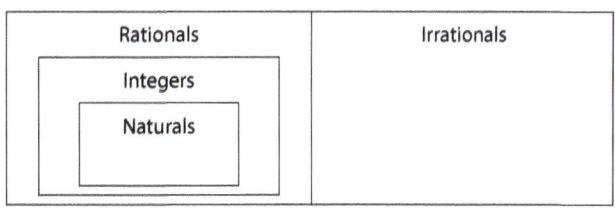

Decimals do not seem to feature in the list above – are they rational or irrational?

- *Recurring decimals* can always be written as fractions so they are rational numbers.
- *Terminating decimals* can also be written as fractions, so they are rational numbers too.
- *Non-recurring, non-terminating decimals* (ie they carry on for ever and never repeat) can never be written as an exact fraction, so they are irrational numbers.

Exact values: $\sqrt{4} = 2$ since 4 is a square number. However, $\sqrt{10}$ cannot be written exactly, like the majority of square roots. It is 3.16228... (the dots indicating that the decimal places continue without recurring). To 4 significant figures $\sqrt{10} = 3.162$, but what do you do if the question asks you to give an *exact* value? The answer is to use square root notation:

$$x^2 = 10 \implies x = \pm\sqrt{10}$$

and this (or other equivalent surds) is the only exact way to write down the solution. And, especially if this is an intermediate answer to a question, it is usually better for calculation purposes.

Example: Find the lengths of a and b in the diagram, giving your answers in an exact form.

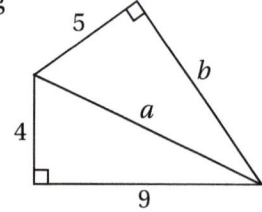

Solution: $a^2 = 9^2 + 4^2 = 97 \Rightarrow a = \sqrt{97}$

$b^2 = a^2 - 5^2 = 97 - 25 = 72$

So $b = \sqrt{72}$

The calculation would have been longer (and possibly less accurate) if we had worked out $\sqrt{97}$ as a decimal and used that.

1.2 Accuracy and Standard Form

When answering questions which have numerical solutions, it is important to understand how to round to an appropriate level of accuracy. And for very large or very small numbers, it is necessary to use standard form.

Accuracy: If there are 6 people in a room, then 6 is perfectly accurate. However, a length given as 6 cm implies that it lies between 5.5 cm and 6.5 cm, and these values are known as the *lower* and *upper bounds*. In theory the upper bound of 6 cm is $6.4\dot{9}$ (ie 6.49999999…), so we should be writing:

$5.5 \leq 6 < 6.5$ (*note the two different symbols*)

It's also important to realise that the number 6.0 implies a greater accuracy:

$5.95 \leq 6.0 < 6.05$

In other words, 6 and 6.0 are different numbers when you consider the range of values each represents.

You'll find more detail about significant figures in my blog at peakib.com

Questions often ask you to answer to a particular number of *significant figures* or *decimal places*. The first significant figure is the first non-zero digit. The first decimal place is the first figure after the decimal point. Try these:

Write the following numbers to 3SF: 41.26, 2096, 21.04, 699.8

Write the following numbers to 1DP: 12.392, 0.061, 4.952

Answers

41.3, 2100, 21.0, 700

12.4, 0.1, 5.0

But when you are working through a calculation, you should *not* round off at intermediate stages. Keep full calculator accuracy until you get to the answer, then round.

If a question asks you to answer to a "suitable" degree of accuracy that usually means you should not be *too* accurate. For example, if a diagram gives lengths to 2SF, your answer should also be to 2SF.

Standard form: Standard form gives us an alternative way of writing very large and very small numbers without using lots of zeroes. For example:

$$43\,000 = 4.3 \times 10\,000 = 4.3 \times 10^4$$

$$23\,000\,000 = 2.3 \times 10\,000\,000 = 2.3 \times 10^7$$

$$0.000\,56 = \frac{5.6}{10\,000} = 5.6 \times \frac{1}{10\,000} = 5.6 \times 10^{-4}$$

$$0.000\,000\,109 = \frac{1.09}{10\,000\,000} = 1.09 \times 10^{-7}$$

Ordinary size numbers can also be written in standard form. For example, $12.5 = 1.25 \times 10^1$. Why would you want to do this? Sometimes calculations involving very big and very small numbers can end up as ordinary size numbers! For example:

$$(2.4 \times 10^8) \times (3 \times 10^{-9}) = 7.2 \times 10^{-1} = 0.72$$

It is important that the first part of the number is between 1 and 10. If you do a calculation and the answer comes out as 12×10^4 this is not in standard form: it must be written as 1.2×10^5.

Some general points about standard form:

- A common mistake is to write eg 4.1×10^3 as 4.1^3
- Make sure you know how to enter numbers in standard form on your GDC, and also how to set the GDC to give answers in standard form. But be careful not to use "calculator notation"; for example, some calculators might display 3×10^{12} as 3E12.
- In exam questions, you may be asked to "give your answer in the form $a \times 10^k$, where $1 \le a < 10$ and $k \in \mathbb{Z}$" – this is just a formal way of defining standard form.

Percentage error: When using a ruler, it's not really possible to measure to more accuracy that 0.5 mm, or 0.05 cm. The error in a measurement is known as the *absolute error*; but if a measurement is just 1.6 cm, say, then an error of 0.05 cm is clearly of greater significance than if the measurement were 10.8 cm. It is useful to express the error as a percentage of the measurement, and this is known as the *percentage error*. Calculate it as $\dfrac{|\text{error}|}{\text{measurement}} \times 100$. So: 1.6 ± 0.05 cm leads to a maximum percentage error of $\dfrac{0.05}{1.6} \times 100 = 3.125\%$ whereas 10.8 ± 0.05 cm leads to a maximum percentage error of $\dfrac{0.05}{10.8} \times 100 = 0.463\%$.

> Go to the website to watch a video solution of a percentage error question.

Example: John has a hemispherical paperweight and he measures its radius as 4.3 cm. He uses this value to calculate its volume.

 (a) Find the volume of the paperweight that John calculated.

 (b) Given that the measurement was to 1 decimal place, find the upper bound of the volume.

 (c) Hence find the maximum percentage error in John's calculation, giving your answer to 3SF.

Solution: (a) $V = \frac{1}{2} \times \frac{4}{3}\pi r^3 = \frac{2}{3}\pi \times 4.3^3 = 166.519...$ cm³.

 (b) Upper bound $= \frac{2}{3}\pi \times 4.35^3 = 172.396...$ cm³.

 (c) Max % error $= \dfrac{|172.396 - 166.519|}{166.519} \times 100 = 3.53\%$

Given that $h = \dfrac{e^x - 2x}{\sqrt{y-z}}$,

 (a) Calculate the value of h when $x = 4.35$, $y = 0.076$ and $z = 0.065$. Write down the full calculator display.

 (b) Write down your answer to part (a)

  (i) correct to two significant figures

  (ii) correct to three decimal places

  (iii) in the form $a \times 10^k$, where $1 \le a < 10$, $k \in \mathbb{Z}$.

(a) 655.7769135	(a) *Take great care with the initial calculation to ensure the answers in part (b) gain full marks.*
(b) (i) 660	
(ii) 655.777	(b) (i) *A rather sneaky one. Make sure you understand the answer.*
(iii) 6.557769135×10^2	

Answers

2.856

15.3%

2205

15.624

$285.71

A reminder about percentages: you should be able to do the following types of calculation:

- What is 12% of 23.8?
- What is 56 as a percentage of 365?
- Increase 2100 by 5%.
- Decrease 18.6 by 16%.
- A coat costs $300 including 5% sales tax. What is the cost of the coat before tax?

Estimation: We are used to making estimates in everyday life, particularly of measures such as length, weight, time and money. Estimates can be used in questions in a variety of ways.

Example: If we estimate that the sun is 1.5×10^8 km from the Earth, and that light travels at 3×10^8 ms^{-1}, find an estimate, in minutes, for the time it takes light to travel from the sun to the Earth, giving your answer to an appropriate degree of accuracy.

Solution: First we must use consistent units. The distance from the sun to the Earth is 1.5×10^{11} m. Next we calculate the time, and the answer at this stage will be in seconds (since the speed is in ms^{-1}).

Always consider the accuracy to which you give an estimate – unless, of course, the question *tells* you what accuracy to use.

Time $= \dfrac{1.5 \times 10^{11}}{3 \times 10^8} = 5 \times 10^2$ s. Then 500 s $= 8.333...$ minutes. Clearly this is too accurate given that the original figures were estimates, so a sensible answer would be "about 8 minutes."

You will often find estimates used in questions involving scatter diagrams and regression lines, since the equation of the line only gives an approximate relationship between the two variables. Other questions may ask you to compare an estimated value with an actual value, perhaps to find the percentage error in the estimate.

Answers

1. (a) 23.4 (b) 1.41 (c) 3260 (d) 0.0878
2. (a) 1.103 (b) 0.008 (c) 12.150 (d) 0.721
3. (a) 1.62×10^6 (b) 8.4×10^{-1} (c) 1.625×10^9
4. 0.0402%
5. 1.296×10^9; 2×10^{12}
6. (a) $16\frac{2}{3}$ ms^{-1} (b) 4.1×10^6 cm^3 (c) $1\,814\,400$ s
7. 0.9 cm^2

Note that if you are not told whether or not to use standard form, just use whatever is convenient.

Accuracy and Standard Form: Practice Exercise

1. Write the following numbers correct to 3 significant figures:

   (a) 23.419   (b) 1.4072   (c) 3255   (d) 0.087848

2. Write the following numbers to 3 decimal places:

   (a) 1.10306   (b) 0.00812   (c) 12.1495   (d) 0.7205

3. Calculate the following, giving answers in standard form:

   (a) 12000×135 (b) 0.0056×150 (c) $(1.3 \times 10^6) \div (8 \times 10^{-4})$

4. Find the percentage error when the fraction $\frac{22}{7}$ is used as an approximation to π. Answer to 4 decimal places.

5. $x = 3.6 \times 10^4$ and $y = 1.8 \times 10^{-8}$.

   Calculate the values of x^2 and $x \div y$, giving your answers in the form $a \times 10^k$ where $1 \le a < 10$ and $k \in \mathbb{Z}$.

6. Convert:

    (a) $60\,\text{km}\,\text{h}^{-1}$ to ms^{-1} (b) $4.1\,\text{m}^3$ to cm^3 (c)   3 weeks to seconds

7. The diameter of a circle is measured as 2.6 cm to 1 decimal place. Calculate the area of a 60° sector of the circle, giving your answer to an appropriate degree of accuracy.

1.3 Sequences and Series

There are many different types of number sequence. You only need to know about two: the *arithmetic sequence* (or *progression*) (AP) and the *geometric sequence* (GP). In an AP each number is the previous number *plus* a constant. In a GP each number is the previous number *multiplied* by a constant. A *series* is the same as a *sequence* except that the terms are added together: thus a series has a *sum*, whereas a sequence does not.

To answer most sequences and series questions, make sure you are familiar with the formulae below. First, the notation:

$$u_1 = \text{the first term of the sequence}$$

$$n = \text{the number of terms in the sequence}$$

$$d = \text{the common difference (the number added on in an AP)}$$

$$r = \text{the common ratio (the multiplier in a GP)}$$

$$u_n = \text{the value of the } n\text{th term}$$

$$S_n = \text{the sum of the first } n \text{ terms}$$

$$S_\infty = \text{the sum to infinity}$$

Try working out the values of d and r in the sequences in the examples box above.

The formulae

For an AP:

The value of the nth term: $u_n = u_1 + (n-1)d$

$$d = u_{n+1} - u_n$$

The sum of n terms: $S_n = \frac{n}{2}(u_1 + u_n) = \frac{n}{2}(2u_1 + (n-1)d)$

For a GP:

The value of the nth term: $u_n = u_1 r^{n-1}$

$$r = \frac{u_{n+1}}{u_n}$$

The sum of n terms: $S_n = \frac{u_1(r^n - 1)}{r - 1}$ or $\frac{u_1(1 - r^n)}{1 - r}$

And for GPs only there is a formula for "the sum to infinity." If the common ratio is in the range $-1 < r < 1$ then the terms get ever smaller and approach zero. In this case, the sum of the series will converge on a particular value. To calculate this value:

The sum to infinity: $S_\infty = \frac{u_1}{1 - r}$

Examples

Arithmetic sequences:

3, 5, 7, 9 …

1.1, 1.3, 1.5, 1.7 …

11, 7, 3, –1, –5 …

Geometric sequences:

1, 3, 9, 27 …

4, 6, 9, 13.5 …

12, 6, 3, 1.5, 0.75 …

2, –6, 18, –54 …

Answers

d: 2, 0.2, –4

r: 3, 1.5, 0.5, –3

The sum formulae always start from the first term. If you wanted to sum, say, the 10th term to the 20th term, you would calculate $S_{20} - S_9$. Think about it!

📱 You should be able to use your calculator for the majority of sequences and series questions. Make sure you are familiar with all the techniques you might need.

Series questions often involve algebra as well as numbers. Note that to find d given two consecutive terms in an AP, subtract the first from the second; and to find r in a GP, divide the second by the first. Let's try working through an exam-style question:

(a)   In an arithmetic sequence, $u_2 = 30$ and $u_6 = 90$. Calculate the common difference and the first term.

Although you could use the formulae, it's sometimes easier to use your fingers! My fingers tell me that you must add on 4 common differences to get from the 2nd term to the 6th.

$$4d = 60 \implies d = 15$$

So we can now subtract the common difference from the second term to get the first term.

$$u_1 = 30 - 15 = 15$$

(b)   The first, second and fourth terms of the arithmetic sequence are the first three terms of a geometric sequence. Find the sum of the first seven terms of the geometric sequence.

Let's see what the three terms are: 15, 30, 60. OK, to calculate the sum of a GP we need the first term, the common ratio and the number of terms.

$$u_1 = 15, r = 2, n = 7.$$

$$S_n = \frac{u_1(r^n - 1)}{r - 1}$$

$$S_7 = \frac{15(2^7 - 1)}{2 - 1} = 1905$$

As a general rule, if you are going to use a formula – write it down first. Then substitute the numbers, and calculate. It shows the examiner what you are doing; and you are less likely to make a mistake.

An arithmetic sequence has third term 12 and sixth term 18.

(a)   Find the common difference.

(b)   Find the first term.

(c)   Find the sum of the first 11 terms.

Using the formula for nth term, we get:

$$u_3 = u_1 + 2d = 12$$

$$u_6 = u_1 + 5d = 18$$

(a)   Solving simultaneously, $3d = 6$ giving $d = 2$.

(b)   Substituting d into the first equation,

$$u_1 + 2 \times 2 = 12 \text{ giving } u_1 = 8.$$

(c) $S_{11} = \frac{11}{2}(2 \times 8 + (11 - 1) \times 2)$

$$= \frac{11}{2}(16 + 20) = 11 \times 18 = 198$$

This has similarities to the last question, but this time I've shown you how to use the formulae rather than fingers.

Once we know the first term and the common difference, we can solve any further questions about the sequence.

It often happens that, where formulae are involved, you are given the value of the variable on the left-hand side, and have to find the value of one of the other variables. This will always involve solving an equation, and the sequence and series formulae are no different in this respect from any others.

Example: Find the number of terms in the geometric series

$$1 + 3 + 9 + 27 + \ldots\ldots + 59\,049$$

Solution: Let's write down the formula for the nth term of a geometric series and then substitute the values we know:

$$u_n = u_1 r^{n-1}$$

$$59\,049 = 1 \times 3^{n-1}$$

$$n = 11 \text{ (GDC)}$$

Remember that the common difference and the common ratio can also be negative:

AP with $u_1 = 7$ and $d = -2$:   7, 5, 3, 1, −1, −3, …

GP with $u_1 = 2$ and $r = -1.5$:   2, −3, 4.5, −6.75, …

Sequences and series questions often result in simultaneous equations, the unknowns normally being the first term and common difference or ratio.

> The sum of the first, second and third terms of a geometric sequence is 19 and the sum to infinity is 27. Find the common ratio and the first term.
>
> $r = \frac{2}{3}, u_1 = 9$

Ways of defining sequences and series: Other than listing the numbers, there are two methods for defining a sequence:

- *Function definition*, eg $u_n = 2n^2 + 1$, giving 3, 9, 19, 33, …
- *Recursive definition*, eg $u_1 = 5$, $u_{n+1} = 1 + \frac{1}{u_n}$ giving $\frac{6}{5}, \frac{11}{6}, \frac{17}{11}, \ldots$

> Note that neither of these sequences is an AP or a GP. *Any* set of numbers forms a sequence.

Sigma notation can be used to turn the function definition of a sequence into a series; in other words, where the terms of the sequence are added together. In the following example, the sigma symbol means "the sum of."

- *Sigma notation* $\sum_1^4 (k^2 - 2) = (1^2 - 2) + (2^2 - 2) + (3^2 - 2) + (4^2 - 2) = 22$

Example: Given that $\sum_{k=1}^{k=n} (4k - 1) = 1176$, find the value of n.

Solution: The series begins $3 + 7 + 11 + 15$, so we are dealing with an AP with $u_1 = 3$ and $d = 4$.

Using the sum formula, $1176 = \frac{n}{2}(6 + (n-1)4)$. Solve with GDC to get $n = 24$.

Sequences and Series: Practice Exercise

These questions give you some practice in the basics – they are not intended to be the equivalent of exam-style questions.

1.   Find the 10th term and the sum of the first 24 terms of the sequence which begins 3, 8, 13, 18 …

Answers

1. 48, 1452
2. 20.25
3. 3
4. 9
5. 16
6. 4, −1
7. −22
8. $\sum_1^{100}(3k - 2)$
9. 20°

2. Find the sum to infinity of a geometric series having a second term of -9 and a fifth term of $\frac{1}{3}$.

3. A geometric series has first term 3.5 and sum to 10 terms 103 334. Find the common ratio.

4. In an arithmetic sequence the first term is 3 and the sixth term is four times greater than the second term. Find the common difference.

5. In a geometric series, the first term is 0.4 and the common ratio is 2.5. Find the least value of n such that $S_n > 300\,000$.

    Not the easiest of questions – perhaps use the table functionality of your GDC?

6. A sequence is defined by $u_n = an + b$. Given that $u_1 = 3$ and $S_3 = 21$, find the values of a and b.

7. Find the value of $\displaystyle\sum_{k=1}^{k=5}(-1)^k 2^k$.

8. Express the series $1 + 4 + 7 + 10 + ... + 298$ in the form $\displaystyle\sum_{1}^{n} f(k)$

9. A circular disc is cut into 12 sectors whose angles are in an arithmetic sequence. The angle of the largest sector is twice the angle of the smallest sector. Find the size of the angle of the smallest sector.

1.4 Sequences and Series: Applications

One important application of sequences and series is their use in solving financial problems involving interest. If a sum of money is invested, the interest is the amount (expressed as a %) that it earns during each period (usually, but not necessarily, a year).

It is important to remember the single calculations which will:

- Find a percentage of an amount
- Increase or decrease an amount by a percentage

    5 % of 650 = 0.05 × 650

    12 % of 2000 = 0.12 × 2000

    Increase 120 by 5 % = 120 × 1.05

    Increase 4500 by 12 % = 4500 × 1.12

    Decrease 55 by 2 % = 55 × 0.98

Simple interest: the interest earned is not added to the total amount which thus stays constant.

- $1000 at 5% simple interest per year will earn $50/year. In 10 years, the investment is worth 1000 + 10 × 50 = $1500.

Compound interest: the interest earned is added to the amount invested. Thus the investment grows by a larger amount each year.

- $1000 at 5% compound interest will multiply by 1.05 each year

    After 1 year, the investment is worth 1000 × 1.05 = $1050

    After 2 years, the investment is worth 1050 × 1.05^2 = $1102.50

    After n years, the investment is worth 1000 × 1.05^n

Note that with simple interest, the value of the investment is increased by $50 each year and will form an AP. With compound interest, the value will multiply by 1.05 each year and will form a GP.

> Beware of questions where extra money is added to the investment each year *as well as* the interest.

Mary invested $2000 for 4 years at 3% simple interest. Mo invested $2000 for 3 years at 4% compound interest. Which investment was then worth more?	
Mary: 3% of $2000 = $60. So after 4 years Mary's investment was worth $2240. Mo: After 3 years Mo's investment was worth $2000 \times 1.04^3 = \$2249.73$. Mo's investment was worth more.	*Sometimes a question asks how much interest the investment made. So in this case Mary's interest totalled $240. Mo's interest totalled $249.73.*

Sometimes a sneaky question uses a time period other than a year, but the calculations are done in the same way. If an investment attracts compound interest of 6% per annum, paid every month, then we can regard this as being the same as 0.5% per month. To find the value of an investment after 6 years, the multiplier would be 1.005^{72} (72 months at 0.5% per month).

Depreciation: Investments attracting interest appreciate in value, whereas some items will lose value. For example, if a car was bought new for €13000 and then depreciated by 30% each year, after three years it would be worth $13000 \times 0.7^3 = €4459$. Radioactive decay provides another example of a decreasing quantity.

Example: A radioactive substance decays such that it loses 4% of its mass per year. After how many full years would 10 kg first reduce to less than 5 kg?

Solution: We need to solve $10 \times 0.96^t < 5$.   First solve $10 \times 0.96^t = 5$ which gives $t = 16.9\ldots$. Thus t will be either 16 or 17: try substituting both to decide.

$$10 \times 0.96^{16} = 5.20\ldots$$

$$10 \times 0.96^{17} = 4.99\ldots$$

Thus $t = 17$ years, since mass is lost each year.

The previous example showed a non-financial application. Here's another one. Note that the question doesn't mention sequences and series – you have to be alert to the fact that this is the area of the syllabus it relates to.

Each day a runner trains for a 9 km race. On the 1st day she runs 1000 m, and then increases the distance by 250 m on each subsequent day.

(a) On which day does she run a distance of 9 km?

(b) What is the total distance in km she will have run in training by the end of that day?

(a) $u_n = u_1 + (n-1)d$ $9000 = 1000 + (n-1) \times 250$ $8000 = 250n - 250$ $\therefore \quad n = 33$ days (b) $S_n = \frac{n}{2}(u_1 + u_n)$ $S_n = \frac{33}{2}(1000 + 9000) = 165\,000$ She will have run $165\,000$ m $= 165$ km	*In part (a) we are dealing with the terms of an AP. We know the value of the last term, but need the number of terms. Notice that the question has distances in both km and m.* *In part (b) we are looking for the sum of the series. It's easier to use the formula containing u_n (the last term) since we know it is 9000.* *Don't forget to convert back to km at the end.*

Amortization and annuities: *Amortization* is the process whereby a loan is paid off using regular payments. The *present value* is the amount of the loan, the *future value* is 0. Use the TVM functionality on your GDC to enter the given values, and then find the "missing" value.

Examples:

Loan amount:	$16500
Interest rate:	8% p.a.
Loan period:	5 years
Payments made:	Quarterly
No. of payments:	$5 \times 4 = 20$
Calculate quarterly payment:	$1009

Loan amount:	€9000
Loan period:	3 years
Payments made:	Monthly
No. of payments:	$3 \times 12 = 36$
Payment amount:	€303
Calculate interest rate:	12.9% p.a.

The total amount to be paid can be calculated by multiplying the payment amount by the number of payments. In the first example this would be $20 180, and in the second €10 908. Note that interest is an annual rate, compounded over the relevant time period. For example, a 12% rate compounded monthly means that each month the amount of interest added to the loan will be 1% of the total remaining to be paid.

The calculations for an *annuity* are pretty much the same. Here, you invest a sum of money (the present value) which is assumed to grow at a constant interest rate. You then take out a fixed amount of cash every year, quarter or month until there is nothing left. Use the TVM app to calculate the amount we should invest so as to receive an annual payment of $1200 for 10 years with an interest rate of 9%. You should find that the initial investment needs to be just over $7700.

If you wanted $100 per month, you would need to invest slightly more. Why?

1.5 Exponents

What are exponents? *Exponent* is another word for power or index. You must understand the meaning of positive and negative whole number powers. You must also be very familiar with the rules for using powers.

First, let's look at powers of 2:

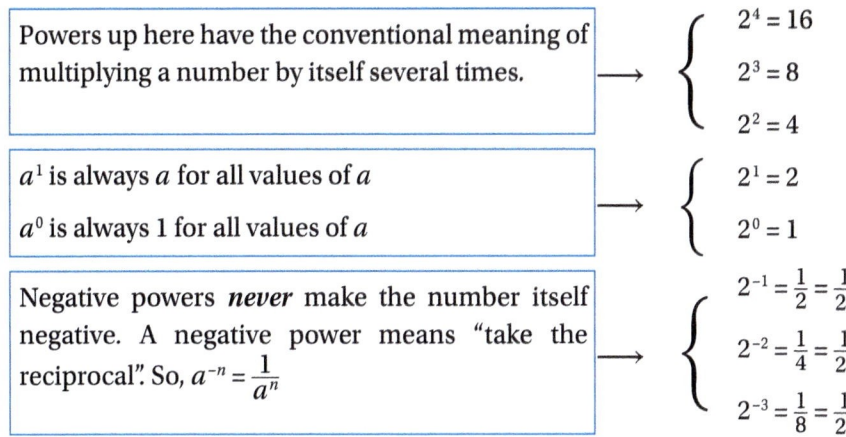

Powers up here have the conventional meaning of multiplying a number by itself several times.	$2^4 = 16$ $2^3 = 8$ $2^2 = 4$
a^1 is always a for all values of a a^0 is always 1 for all values of a	$2^1 = 2$ $2^0 = 1$
Negative powers *never* make the number itself negative. A negative power means "take the reciprocal". So, $a^{-n} = \dfrac{1}{a^n}$	$2^{-1} = \frac{1}{2} = \frac{1}{2^1}$ $2^{-2} = \frac{1}{4} = \frac{1}{2^2}$ $2^{-3} = \frac{1}{8} = \frac{1}{2^3}$

Fractional powers always involve *roots*.

Power	Meaning
$x^{\frac{1}{2}}$	$\sqrt{x}$
$x^{\frac{1}{3}}$	$\sqrt[3]{x}$
$x^{\frac{3}{2}}$	$\sqrt{x^3} = (\sqrt{x})^3$
$x^{-\frac{2}{5}}$	$\dfrac{1}{\sqrt[5]{x^2}}$

Examples:

$2.5^1 = 2.5$

$4^{-2} = \dfrac{1}{16}$

$\left(\dfrac{2}{3}\right)^{-3} = \left(\dfrac{3^3}{2^3}\right) = \dfrac{27}{8}$

$8^{\frac{5}{3}} = \left(\sqrt[3]{8}\right)^5 = 32$

In general, $x^{\frac{1}{n}} = \sqrt[n]{x}$ and $x^{\frac{m}{n}} = \sqrt[n]{x^m} = \left(\sqrt[n]{x}\right)^m$.

Laws of exponents: The rules which follow occur in all sorts of mathematical situations and you should learn them carefully:

Examples

- $a^x \times a^y = a^{x+y}$ $2^{x+3} = 2^x \times 2^3 = 8 \times 2^x$

- $a^x \div a^y = a^{x-y}$ $\dfrac{x^3}{x^5} = x^{3-5} = x^{-2}$ or $\dfrac{1}{x^2}$

- $(a^x)^y = a^{xy}$ $9^x = (3^2)^x = 3^{2x}$

- $(ab)^x = a^x b^x$ $(3x)^3 = 3^3 \times x^3 = 27x^3$

Simplifying expressions containing exponents: Fractional powers always involve roots, and converting roots to powers helps to simplify expressions. For example, we can use $9 = 3^2$ to simplify $\dfrac{\sqrt{3} \times 9^{\frac{2}{3}}}{\sqrt[3]{9}}$:

$$\frac{\sqrt{3} \times 9^{\frac{2}{3}}}{\sqrt[3]{9}} = \frac{3^{\frac{1}{2}} \times (3^2)^{\frac{2}{3}}}{(3^2)^{\frac{1}{3}}} = \frac{3^{\frac{1}{2}} \times 3^{\frac{4}{3}}}{3^{\frac{2}{3}}} = \frac{3^{\frac{11}{6}}}{3^{\frac{2}{3}}} = 3^{\frac{7}{6}}$$

You may also have to use the laws of exponents to simplify algebraic expressions. For example, simplify $\dfrac{6\sqrt{x} \times 3x}{9x^{\frac{5}{2}}}$:

$$\frac{6\sqrt{x} \times 3x}{9x^{\frac{5}{2}}} = \frac{6x^{\frac{1}{2}} \times 3x^{1}}{9x^{\frac{5}{2}}} = \frac{18x^{\frac{3}{2}}}{9x^{\frac{5}{2}}} = \frac{2}{x}$$

The next exam-style question is obviously about exponents but, look carefully, and it's actually a quadratic equation. In other words, it's of the form something^2 + something + number. You will also find this form when dealing with trigonometric equations and equations involving e^x.

(a) Show that the equation $(2^x)^2 + 2^x - 20 = 0$ can be written in the form $(2^x + a)(2^x + b) = 0$, where a and b are integers to be found.

(b) Hence solve the equation for x and explain why there is only one solution.

(a) $(2^x)^2 + 2^x - 20 = 0$ $y^2 + y - 20 = 0$ $(y + 5)(y - 4) = 0$ $(2^x + 5)(2^x - 4) = 0$ (b) So $2^x = -5$ or $2^x = 4$. But exponentials cannot take negative values, so the only solution is $2^x = 4$, giving $x = 2$.	*It may help to use y as the base of quadratic, then replace y with 2^x.* *When solving equations in exams, it's always a good idea to substitute your answer into the original equation to check you are right. In this case, with x = 2, we get* $4^2 + 4 - 20 = 16 + 4 - 20 = 0$

Surds: Make sure you are entirely familiar with the rules for manipulating surds. In particular:

$$\sqrt{a} \times \sqrt{b} = \sqrt{ab}, \ \frac{\sqrt{a}}{\sqrt{b}} = \sqrt{\frac{a}{b}}$$

$$\sqrt{a} + \sqrt{b} \neq \sqrt{a + b}, \ \sqrt{a} - \sqrt{b} \neq \sqrt{a - b}$$

Thus: $\sqrt{2} \times \sqrt{18} = \sqrt{36} = 6$

$\sqrt{50} + \sqrt{98} = \sqrt{25}\sqrt{2} + \sqrt{49}\sqrt{2} = 5\sqrt{2} + 7\sqrt{2} = 12\sqrt{2}$

$\sqrt{\dfrac{49}{100}} = \dfrac{\sqrt{49}}{\sqrt{100}} = \dfrac{7}{10}$

It is sometimes useful when manipulating surds to think of the rules of algebra – in the second example above, compare $5\sqrt{2} + 7\sqrt{2}$ with $5x + 7x$. This should enable you to quickly calculate the values of $(4\sqrt{3})^2$ and $(2 + \sqrt{5})(2 - \sqrt{5})$. (Answers: 48 and –1)

You should also be able to *rationalise* the denominators of fractions which contain surds. If the denominator is of the form $\sqrt{a}$, then multiply numerator and denominator by $\sqrt{a}$; if the denominator is of the form $a + \sqrt{b}$, then multiply numerator and denominator by $a - \sqrt{b}$. In both cases, the denominator will become a rational number. (The process is very similar to the division of complex numbers.)

For example: $\dfrac{4}{3 - \sqrt{3}} = \dfrac{4}{3 - \sqrt{3}} \times \dfrac{3 + \sqrt{3}}{3 + \sqrt{3}} = \dfrac{12 + 4\sqrt{3}}{9 - 3} = \dfrac{12 + 4\sqrt{3}}{6} = 2 + \dfrac{2\sqrt{3}}{3}$

1.6 Logarithms

What is a logarithm? The mapping diagram in the notes box shows the function $f(x) = 2^x$ applied to a few integers. The inverse of this function would map $8 \to 3$, $4 \to 2$ and so on: in other words, it would find what power of 2 gives the required number. As shown at the bottom of the diagram, the inverse is the logarithm function. So the logarithm to the *base 2* of a number is the power of 2 which gives the number. For example, $\log_2 16 = 4$. It may be helpful to think of the relationship of the log(arithm) function to the power function as similar to that between the square root function and the square function.

Mapping $x \to 2^x$
gives $3 \to 8$
           $2 \to 4$
           $1 \to 2$
           $0 \to 1$
           $-1 \to 0.5$
and $\log_2 x \leftarrow x$

Laws of logarithms: Because logarithms are just powers, the laws of logarithms are very similar to the laws of exponents. You should be very familiar with them, although you will find them in the formula book. These rules apply to logs with any base.

- $\log a + \log b = \log(ab)$
- $\log a - \log b = \log(a/b)$
- $\log a^n = n \log a$

Examples:

$$\log_3 27 = 3$$
$$\log_{10} 0.1 = -1$$
$$\log_2(\sqrt{2}) = \tfrac{1}{2}$$
$$\log_5(5^x) = x$$

I used base 2 to illustrate what a logarithm is. In fact, you only need to know about logs to the base 10 and to the base e (the number e is explained in more detail in the next chapter on functions).

So the important thing to remember is that:

- $10^x = y$ is equivalent to $x = \log_{10} y$
- $e^x = y$ is equivalent to $x = \log_e y$, normally written as $\ln y$.

Depending on your calculator, it is likely that it has log (which is $\log_{10}$) and ln keys, with 2nd function 10^x and e^x.

We have seen that negative powers do not lead to negative numbers. It therefore follows that it isn't possible to find the logarithm of a negative number; for example, $\log_{10}(-100)$ would mean "what power of 10 gives -100". And the answer is: not possible!

The formula connecting the pH level of a substance (its acidity or alkalinity) with its hydrogen ion concentration, H^+, in moles per litre, is $pH = -\log_{10} H^+$

(a) Find the pH level of a substance with hydrogen ion concentration of 0.0003 moles/litre.

Substances with a pH level less than 7 are acidic.

(b) Determine if ammonia is acidic if it has an H^+ value of 1.3×10^{-9}.

(c) Calculate the largest value of H^+ for which a substance will be acidic.

(a) $pH = -\log_{10} 0.0003 = 3.52$

(b) $pH = -\log_{10}(3 \times 10^{-9}) = 8.89$

   $8.89 > 7$ so ammonia is not acidic

(c) $7 = -\log_{10} H^+$

   $-7 = \log_{10} H^+$

   $H^+ = 10^{-7}$

   For a substance to be acidic, $H^+ > 10^{-7}$

The first two parts show us that as H^+ decreases, pH increases. That's why in part (c) we are asked for the largest value of H^+: any value larger than this, and the pH will drop below 7.

Note how we can use connection between a log function and an exponent function to rewrite the equation in part (c).

Logarithms can crop up in all sorts of unlikely places. Examiners quite like including them in sequences and series questions.

Example: The first two terms of an infinite geometric series are $3 \ln x$ and $\ln x$. Find the sum of the series.

Solution: We need the common ratio which is $\frac{u_2}{u_1} = \frac{\ln x}{3 \ln x} = \frac{1}{3}$. Now we know it's an infinite sequence so the sum is $\frac{u_1}{1-r} = \frac{3 \ln x}{1 - \frac{1}{3}} = \frac{9 \ln x}{2}$.

No need to use the laws of logarithms there. But how about this one?

Example: The first two terms of an arithmetic sequence are $\log_{10} x$ and $\log_{10}\left(\frac{x}{10}\right)$. Find the common difference, giving your answer as an integer.

Solution: $d = u_2 - u_1 = \log_{10}\left(\frac{x}{10}\right) - \log_{10} x = \log_{10}\left(\frac{x/10}{x}\right) = \log_{10}\frac{1}{10} = -1$

1.7 Polynomial Equations

A *polynomial* is an expression of the form $a + bx + cx^2 + dx^3 + ...$, where a, b, c, d ... are constants. The highest power of x determines the *degree* of the polynomial. Thus:

- $3x^2 - 4x + 1$ is a degree 2 polynomial, also called a *quadratic*
- $x^3 - 4x + 5$ is a degree 3 polynomial, also called a *cubic*

Note that not all terms have to be present. So x^5 on its own is still a degree 5 polynomial.

If a polynomial is written as $P(x)$, then $P(x) = 0$ is a polynomial equation, and your calculator will have an app to solve such equations. In graphical terms, to solve $P(x) = 0$ we need to find where the graph of $P(x)$ intersects the x-axis. Such points are known as *zeroes* of the function, and also as *roots* of the equation. The number of solutions entirely depends on the graph and its relationship to the x-axis.

The graph of a quadratic function $f(x)$ will have a shape like this:

Although when considering complex numbers, **all** quadratic equations have 2 solutions (which includes the case of a "repeated" solution).

Now let's put some axes in, and see how many roots there are to $f(x) = 0$.

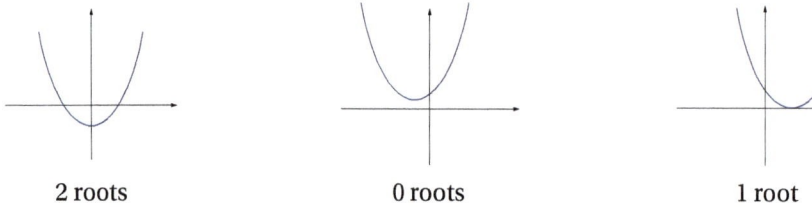

| 2 roots | 0 roots | 1 root |

Similarly, a cubic polynomial may have a shape like this. See if you can align the axes so to show that a cubic equation could have 1, 2, or 3 roots.

If you want to see how your GDC deals with a polynomial equation with no roots, try solving $x^2 + x + 1 = 0$.

When you use your GDC app to solve a polynomial equation, you will need to enter the degree of the

polynomial, and then the *coefficients* – these are the constants. The GDC will then show all the possible solutions.

Note the following points:

- The equation must always be in the form $P(x) = 0$ if you are using the polynomial equation app. This may mean rearranging an equation. For example, $2(x^2 - 3) = 3x - 4$ can be rearranged to become $2x^2 - 3x - 2 = 0$. (Solutions: $x = 2$ or $x = -0.5$)
- If there are missing terms, you must put in coefficients equal to 0. For example, the coefficient of x^2 is zero in the equation $x^3 - 3x + 2 = 0$. (Solutions: $x = -2$ or $x = 1$)

Example: Differentiate the function $f(x) = x^4 - 3x^2 + 1$ and hence find the x coordinates of any stationary points, giving answers to 3 decimal places.

Solution: $f'(x) = 4x^3 - 6x$

For stationary points, $f'(x) = 0 \therefore x = -1.225, 0, 1.225$

> Differentiation and turning points are dealt with in chapter 5 – see page 179.

1.8 Complex Numbers

The square root of -1: The term "imaginary number" is off-putting because it makes $\sqrt{-1}$ seem very abstract. Remember that negative numbers do not exist in real life either: you cannot have a negative amount of anything. As with negative numbers, once $\sqrt{-1}$ has been defined, we can fit it into our existing algebra – and very useful it turns out to be.

$$x^2 + 2x + 5 = 0$$
$$x = \frac{-2 \pm \sqrt{4 - 20}}{2}$$
$$x = \frac{-2 \pm \sqrt{-16}}{2}$$
$$= \frac{-2 \pm 4i}{2} \Rightarrow x = -1 \pm 2i$$

Definitions: The notes box shows the solution of $x^2 + 2x + 5 = 0$. The two solutions, $-1 + 2i$ and $-1 - 2i$ are called *complex numbers* and cannot be further simplified. A complex number z has the form $a + ib$, where a is called the *real part* and b is the *imaginary part*. These can be denoted by $\operatorname{Re} z$ and $\operatorname{Im} z$. Note that all real numbers can be considered as complex numbers with a zero imaginary part. The *conjugate* of $z = a + ib$ is $z^* = a - ib$: thus complex roots of quadratic equations always occur in conjugate pairs.

Basic arithmetic: To *add* complex numbers, add their real parts together, then add their imaginary parts.

$$(a + ib) + (c + id) = (a + c) + i(b + d)$$

Example: $(2 + 3i) + (4 - i) = 6 + 2i$

Subtraction is similar:

$$(a + ib) - (c + id) = (a - c) + i(b - d)$$

Example: $(4 + 6i) - (5 + 3i) = -1 + 3i$

Multiplication uses the expansion of brackets:

$$(a + ib) \times (c + id) = ac + iad + ibc + i^2bd = (ac - bd) + i(ad + bc)$$

Example: $(2 + 4i) \times (2 + 3i) = 4 + 6i + 8i + 12i^2 = -8 + 14i$

> Your GDC has an i button, so you can do all complex number calculations. Set in complex mode, it will also give complex solutions to equations.

> Never forget that $i^2 = -1$

One useful fact that follows is that both the addition and multiplication of conjugate pairs give real results.

$$(a + ib) + (a - ib) = 2a \text{ (or } z + z^* = 2\operatorname{Re}(z))$$

$$(a + ib) \times (a - ib) = a^2 - i^2b^2 = a^2 + b^2 \text{ (or } zz^* = |z|^2)$$

The second result leads to the method for division of complex numbers – similar to the rationalisation of surds.

To *divide* complex numbers:

1. Write the division as a fraction

2. Multiply top and bottom by the conjugate of the bottom

3. Simplify to the form $a + ib$

Example: Divide $3 + 2i$ by $1 - 4i$

$$\frac{3 + 2i}{1 - 4i} = \frac{(3 + 2i)(1 + 4i)}{(1 - 4i)(1 + 4i)} = \frac{-5 + 14i}{17} = \frac{-5}{17} + \frac{14}{17}i$$

Equality of complex numbers: If two complex numbers are equal then their real parts can be equated, and their imaginary parts can be equated, often resulting in two equations for the price of one. Use this to solve equations involving complex numbers.

Example: Solve $z^2 = 21 + 20i$

See website for the full working

Solution: It often helps (but not always!) when solving questions involving complex algebra to write z as $a + bi$. So we get:

$$(a + bi)^2 = 21 + 20i \Rightarrow a^2 - b^2 + 2abi = 21 + 20i$$

Now we can equate real and imaginary parts and, since this leads to two equations, we can solve simultaneously.

$$a^2 - b^2 = 21 \text{ and } 2ab = 20 \Rightarrow b = \pm 2 \text{ or } \pm 5i$$

But b is real, so $b = \pm 2$ and $a = \pm 5$ giving $z = 5 + 2i$ or $-5 - 2i$

The complex number z satisfies $\frac{z}{z + 2} = 2 - i$. Find:

(a) z

(b) $\arg(z)$ in an exact form

(c) $|z + 2|$

(a) Let $z = x + iy$

$$\frac{x + iy}{(x + 2) + iy} = 2 - i$$

$$x + iy = (2 - i)\left((x + 2) + iy\right)$$

$$x + iy = 2x + 4 + 2iy - ix - 2i - i^2 y$$

$$x + iy = (2x + y + 4) + i(2y - x - 2)$$

Equating real parts and imaginary parts:

$$x = 2x + y + 4 \Rightarrow x + y = -4$$

$$y = 2y - x - 2 \Rightarrow x - y = -2$$

Solving simultaneously, $x = -3$ and $y = -1$

Thus $z = -3 - i$

(b) $\arg(z) = \pi + \arctan\left(\frac{1}{3}\right)$

(c) $|z + 2| = |-3 - i + 2|$

$$= |-1 - i|$$

$$= \sqrt{2}$$

Why did I use x and y instead of a and b? Why not?!

The most important point to note in the working is that wherever a complex number with several terms has appeared, I have collected together real and imaginary parts – see particularly lines 2 and 5 of part (a).

Parts (b) and (c) relate to modulus-argument form, covered in the next section. But note that in (b) it is important to show the point in an Argand diagram so you can see which quadrant it is in.

Once again, in part (c), I have written z as x + iy and then, in line 2, collected together real and imaginary parts.

Try doing all parts with your GDC as well.

1.9 The Complex Plane

Argand diagrams: We can use a *number line* to interpret real numbers geometrically. Each number is represented by a point on the line, or by a position vector – see the number 2 below.

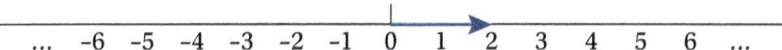

The operation of multiplication by –1 can be represented geometrically by a rotation of 180°. Since $i^2 = -1$, multiplication by i can therefore be represented by a 90° rotation.

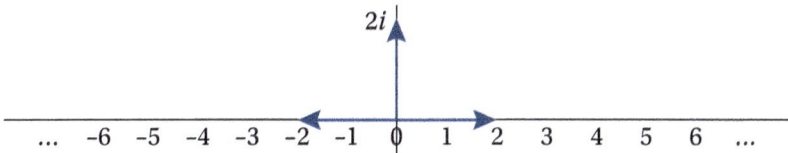

This operation puts $2i$ along a new number line at right angles to the real number line. This is called the *imaginary* axis, and explains why you cannot "see" complex numbers when just operating with real numbers. If we now use the two number lines (real and imaginary) to form two axes, the result is the *complex plane*. Complex numbers can be represented by points or vectors, thus forming *Argand diagrams*.

Modulus of a complex number: The Argand diagram gives us a way of comparing the size of complex numbers to the size of real numbers. Clearly the numbers 2 and –2 have size (or modulus) 2; this is also the length of their vector representations. The vectors representing $2i$ and $3 + 2i$ have lengths 2 and $\sqrt{13}$: using modulus notation we can write $|3 + 2i| = \sqrt{13}$. In general $|a + bi| = \sqrt{a^2 + b^2}$.

> Note that
> $zz^* = (a + bi)(a - bi)$
> which expands to $a^2 + b^2$.
> Thus $zz^* = |z|^2$.

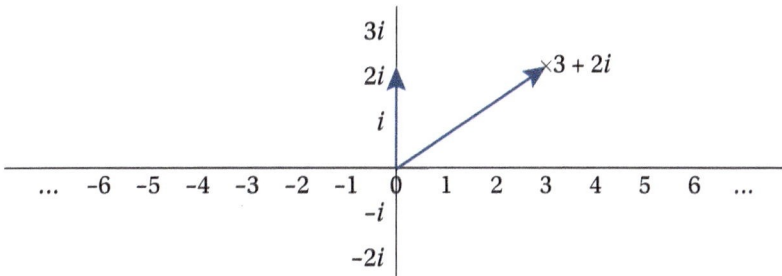

***r*cosθ + *i*r*sinθ form:** Looking at the vector representation above suggests that a complex number could conveniently be written as a length (*modulus*) and angle (*argument*). The argument is the angle made with the real axis, and is usually written such that it lies between $-\pi$ and π (although it can also be expressed in degrees). Thus, $z = 3 + 2i$ has length $\sqrt{13}$ and angle $\arctan(2/3) \approx 33.7°$. Working backwards, this means that $\text{Re}(z) = \sqrt{13} \cos 33.7°$ and $\text{Im}(z) = \sqrt{13} \sin 33.7°$.

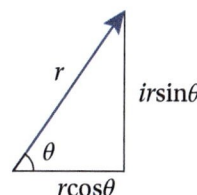

We now have an alternative form for z:

$3 + 2i = \sqrt{13} \cos 33.7° + i\sqrt{13} \sin 33.7° = \sqrt{13} \left(\cos 33.7° + i \sin 33.7° \right)$

This can be written in shorthand as $\sqrt{13}$ cis 33.7°.

Note that conversion from $a + ib$ form to $r\cos\theta + ir\sin\theta$ form is the same as converting Cartesian to polar coordinates - make sure you can do this on your calculator (and the reverse as well).

In summary:

- $|a + ib| = \sqrt{(a^2 + b^2)}$
- $\arg(a + ib) = \tan^{-1}(b/a)$ *if in quadrants 1 or 2, otherwise add* π
- $\mathrm{Re}(r\mathrm{cis}\theta) = r\cos\theta$
- $\mathrm{Im}(r\mathrm{cis}\theta) = r\sin\theta$

An alternative form for $r\mathrm{cis}\theta$ is $re^{i\theta}$. You don't need to know the derivation of this, but must recognise that, for example, the complex number $3e^{\frac{\pi}{3}i}$ has modulus 3 and argument $\frac{\pi}{3}$. An important result that follows is that $re^{i\theta} \times se^{i\phi} = rse^{i(\theta+\phi)}$. In other words, you can multiply complex numbers by multiplying their moduli and adding their arguments. Similarly, $re^{i\theta} \div s^{i\phi} = \frac{r}{s}e^{(\theta-\phi)}$. In $r\mathrm{cis}\theta$ form, these results are:

- $r\mathrm{cis}\theta \times s\mathrm{cis}\phi = rs\mathrm{cis}(\theta + \phi)$
- $r\mathrm{cis}\theta \div s\mathrm{cis}\phi = \frac{r}{s}\mathrm{cis}(\theta - \phi)$

Example: Show that $\dfrac{6 + 2i}{1 + 2i} = \sqrt{8}\, e^{-\frac{i\pi}{4}}$

Solution:
$$\frac{6 + 2i}{1 + 2i} = \frac{(6 + 2i)(1 - 2i)}{(1 + 2i)(1 - 2i)}$$

$$= \frac{10 - 10i}{5} = 2 - 2i$$

$|2 - 2i| = \sqrt{2^2 + 2^2} = \sqrt{8}$, $\arg(2 - 2i) = \arctan(-1) = -\frac{\pi}{4}$

$$\therefore\ 2 - 2i = \sqrt{8}\, e^{-\frac{i\pi}{4}}$$

Complex numbers as vectors: An Argand diagram represents a complex number as a vector. For example, $3 - 4i$ is equivalent to the vector $\begin{pmatrix} 3 \\ -4 \end{pmatrix}$. Addition of complex numbers can be represented by the addition of vectors:

$$(3 - 4i) + (-1 + 6i) = (2 + 2i) \text{ is equivalent to } \begin{pmatrix} 3 \\ -4 \end{pmatrix} + \begin{pmatrix} -1 \\ 6 \end{pmatrix} = \begin{pmatrix} 2 \\ 2 \end{pmatrix}.$$

Multiplication of complex numbers also has a geometric transformational equivalence, but rather less obviously. Multiplying z by the complex number $r\mathrm{cis}\theta$ is equivalent to multiplying $|z|$ by r, and then rotating by θ. Division, of course, will divide $|z|$ by r and rotate by $-\theta$.

Complex Number Calculations: Quick Practice

1. $z = 3 - 2i,\ w = 4 + i$

    Calculate the following by hand, then check on your GDC:

    $$w + z,\ w - z,\ wz,\ \frac{w}{z},\ z^3$$

2. (a) Convert $-1 - \sqrt{3}\, i$ to modulus-argument form by hand.

    (b) Convert $4\,\mathrm{cis}\,\pi$ to $a + bi$ form by hand.

    Check both answers on your GDC.

3. (a) Multiply $\sqrt{2}\, e^{\frac{i\pi}{9}} \times \sqrt{8}\, e^{\frac{i\pi}{9}}$ giving your answer in the same form.

    (b) Repeat part (a) but divide the two numbers.

📟 Find out how to set your GDC to calculate either in $a + bi$ mode or in $re^{i\theta}$ mode. As ever, also check that you are in degrees or radians as appropriate.

The question can be solved on your GDC by entering the division sum using $a + bi$ form, but ensure your GDC is set to $re^{i\theta}$ form for the solution. However, the "show that" means that at least some of this working is necessary.

Answers

1. $7 - i,\ -1 + 3i,\ 14 - 5i,$ $\frac{10}{13} + \frac{11}{13}i,\ -9 - 46i$

2. (a) $2\,\mathrm{cis}(-\frac{2\pi}{3})$ or $2\,\mathrm{cis}(-120)°$

   (b) -4 or $-4 + 0i$

3. $4e^{\frac{5i\pi}{9}},\ \frac{1}{2}e^{-\frac{\pi}{9}}$

The complex number $z = (a - i\sqrt{3})^2$, $a > 0$, has $\text{Re}(z) = -2$.

(a) Find the value of a.

(b) Hence calculate $\frac{1}{z}$ giving your answer in modulus-argument form.

(a) $(a - i\sqrt{3})^2 = a^2 - 2i\sqrt{3} - 3$	(a) Note how to extract the real part of z.
$\text{Re}(z) = a^2 + 3 = 4$	
$\therefore \ a = 1$	(b) Now we know the value of a, the rest of the question can be solved using the GDC.
(b) $\frac{1}{z} = \frac{1}{(1 - i\sqrt{3})^2} = 0.25\,e^{\frac{2\pi i}{3}}$	
$= 0.25\,\text{cis}\,\frac{2\pi}{3}$ (GDC)	

1.10 Adding Sinusoidal Functions

The graphs of functions such as $f(x) = 10\sin(3x + 30)$ are considered further in the chapter on geometry. Such functions are used to model waves, and each of the constants relates to a particular property of the wave. In the example above, the *amplitude* of the wave is 10, the *period* of the wave is $360 \div 3 = 120$, and the 30 represents a *phase shift* of the wave in the horizontal direction. Waves can be added together – for example, two sound waves produce a new sound, or two waves representing the voltage of alternating current can be added to create a new wave. It turns out that, even if there is a phase shift between the waves, addition of sinuisodal waves with the same frequencies produces another sinusoidal wave, and we can use complex numbers to find its equation.

If the two waves have equations $y = a_1\cos(kx + b_1)$ and $y = a_2\cos(kx + b_2)$, then:

1. Create the complex number $z = a_1\text{cis}\,b_1 + a_2\text{cis}\,b_2$

2. The amplitude of the new wave will be $|z|$

3. The phase shift of the new wave will be $\arg(z)$

4. The frequency of the new wave will be unchanged.

For example, the following graph shows the two waves with equations $y = 10\cos(30x + 10)°$ and $y = 20\cos(30x + 30)°$. Both have a frequency of 12, but they have phase shifts of 10° and 30° respectively.

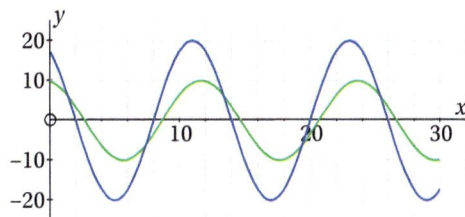

And here is the graph of $y = 10\cos(30x + 10)° + 20\cos(30x + 30)°$:

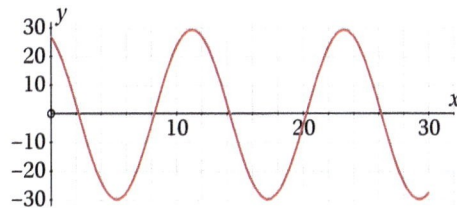

$$z = 10 \operatorname{cis} 10° + 20 \operatorname{cis} 30°$$

$$= (10 \cos 10° + 20 \cos 30°) + i(10 \sin 10° + 20 \sin 30°)$$

$$= 27.17 + 11.74i$$

$$= 29.6 \operatorname{cis} 23.4°$$

Thus the combined wave has equation $y = 29.6\cos(30x + 23.4)°$.

Now show that for the two sinusoidal functions $f(t) = 12 \cos 15t$ and $g(t) = 18 \cos\left(15t - \frac{\pi}{3}\right)$, $f(t) + g(t) = 2\sqrt{171} \cos(15t - 0.639)$.

1.11 Basics of Matrices

A matrix is essentially a table of numbers enclosed in brackets. Most of the matrices you meet will be "square" – they have the same numbers of rows and columns. The numbers in a matrix are called the *elements* of the matrix, and the *order* of a matrix is the number of rows and columns it has. Thus a 2 × 3 matrix has 2 rows and 3 columns.

 It is essential that you are familiar with the matrix features of your GDC. Make sure you can do each of the following:

- Enter and edit matrices of any size
- Find the determinant of a matrix
- Find the inverse of a matrix
- Perform matrix calculations

However, since a question may be asked where some of the elements of a matrix are algebraic, you must also be able to carry out matrix operations without a calculator.

Algebra of matrices: Matrices are usually denoted by capital letters.

- If $A = B$ then all the corresponding elements of the two matrices are the same.
- Matrices can be added or subtracted by adding or subtracting corresponding elements. For example,

$$\begin{pmatrix} 2 & 4 \\ 0 & -1 \end{pmatrix} + \begin{pmatrix} 3 & -4 \\ 2 & 2 \end{pmatrix} = \begin{pmatrix} 5 & 0 \\ 2 & 1 \end{pmatrix}$$

- Multiplying by a scalar will multiply each element by the scalar. eg:

If $A = \begin{pmatrix} 1 & -1 & 3 \\ 0 & 2 & -4 \end{pmatrix}$ then $2A = \begin{pmatrix} 2 & -2 & 6 \\ 0 & 4 & -8 \end{pmatrix}$

| Note that the answer to 1st row × 1st column goes in the $R_1 C_1$ position in the answer matrix, and so on. |

- Multiplication of matrices is a rather curious operation. To multiply two matrices, split the first into rows, the second into columns. Then continue like this:

$$\begin{pmatrix} a & b \\ c & d \end{pmatrix}\begin{pmatrix} p & q \\ r & s \end{pmatrix} = \begin{pmatrix} ap + br & aq + bs \\ cp + dr & cq + ds \end{pmatrix}$$

It is possible to multiply non-square matrices, but only if the number of columns in the first is equal to the number of rows in the second.

eg:

$$\begin{pmatrix} 2 & 0 & -2 \\ 3 & 1 & 2 \end{pmatrix}\begin{pmatrix} 1 & -1 \\ 4 & 2 \\ 0 & 2 \end{pmatrix} = \begin{pmatrix} 2 & -6 \\ 7 & 3 \end{pmatrix}$$

Note that we are multiplying a (2 × 3) by a (3 × 2). The inner numbers show that the multiplication is possible, the outer ones give the size of the answer matrix.

(The lines have only been included in an attempt to be helpful!)

The properties of matrix multiplication can be compared to those of numeric multiplication:

✓ $A(BC) = (AB)C$

✓ $A(B + C) = AB + AC$

but

✗ $AB \neq BA$

The identity matrix: When multiplying numbers, 1 is called the *identity* because $a \times 1 = 1 \times a = a$; that is, it leaves other numbers unchanged. For multiplication, the 2×2 identity matrix is $\begin{pmatrix} 1 & 0 \\ 0 & 1 \end{pmatrix}$ and the 3×3 is $\begin{pmatrix} 1 & 0 & 0 \\ 0 & 1 & 0 \\ 0 & 0 & 1 \end{pmatrix}$. The identity matrix is denoted using *I*. So for any matrix *A*, $AI = IA = A$.

> Here is the identity matrix in action:
> $\begin{pmatrix} 2 & 1 \\ 4 & 3 \end{pmatrix}\begin{pmatrix} 1 & 0 \\ 0 & 1 \end{pmatrix} = \begin{pmatrix} 2 & 1 \\ 4 & 3 \end{pmatrix}$

Zero matrices: In a zero matrix every element is 0. Zero matrices are rather dull! The zero matrix has the symbol *0*, and has the property that for any matrix *A*, $A + 0 = 0 + A = A$.

1.12 Determinants and Inverse Matrices

Determinants: The determinant of a square matrix is a rather useful number associated with it. You can use your GDC to calculate the determinant of any size matrix, but you must also be able to calculate the determinant of a 2×2 matrix by hand.

For a 2×2 matrix $M = \begin{pmatrix} a & b \\ c & d \end{pmatrix}$, $\det M = ad - bc$.

Note that if you multiply two matrices *A* and *B* together, then the determinant of the product $\det AB = \det A \times \det B$.

Inverse of a square matrix: Using numbers as an example again, the inverse of 4 under multiplication is $\frac{1}{4}$ because $4 \times \frac{1}{4} = 1$, the identity. So the inverse of matrix *A* has to be found such that $AA^{-1} = I$. For a 2×2 matrix $\begin{pmatrix} a & b \\ c & d \end{pmatrix}$, the procedure is:

- Work out the determinant
- Swap round *a* and *d*
- Change the signs of *b* and *c*
- Divide by the determinant (either by dividing each element or by putting a fraction in front of the matrix).

Thus, the inverse of $\begin{pmatrix} 2 & 3 \\ 1 & 4 \end{pmatrix}$ is $\frac{1}{5}\begin{pmatrix} 4 & -3 \\ -1 & 2 \end{pmatrix} = \begin{pmatrix} 0.8 & -0.6 \\ -0.2 & 0.4 \end{pmatrix}$.

Since the inverse of a 3×3 is much harder to calculate, you will always be able to use your calculator to find it. However, the principle that $AA^{-1} = I$ is the same as the following question demonstrates.

Example: $M = \begin{pmatrix} 2 & -1 & p \\ 1 & 2 & 0 \\ 2 & 1 & 2 \end{pmatrix}$, $M^{-1} = \frac{1}{2}\begin{pmatrix} -4 & -6 & 8 \\ 2 & 4 & -4 \\ 3 & q & -5 \end{pmatrix}$. Find *p* and *q*.

Solution: We know that $MM^{-1} = I$. But there is no need to multiply out the matrices in full. $R_1 \times C_1 = -4 - 1 + 1.5p = 1$, thus $p = 4$. We could then go for $R_1 \times C_2$ to find *q*. This multiplies out to give $-6 - 2 + \frac{1}{2}pq = 0$, and hence $q = 4$. Also try $R_3 \times C_2$ – check that this too gives us $q = 4$. Alternatively, having found *p*, we could use the GDC to find M^{-1} and hence *q*.

Let $A = \begin{pmatrix} 1 & 4 \\ q & 2 \end{pmatrix}$ and $B = \begin{pmatrix} 7 & p \\ 1 & 2 \end{pmatrix}$, where p and $q \in \mathbb{Z}$ and $p < 10$. Given that $\det A = \det B$ and that $\det AB = 25p$,

(a)   Show that p satisfies the equation $p^2 - 53p + 196 = 0$

(b)   Hence find the values of p and q.

(a) $\det AB = \det A \times \det B$ Since $\det A = \det B$, $\det AB = (\det B)^2$ $\therefore \quad 25p = (14 - p)^2$ $ 25p = 196 - 28p + p^2$ $ p^2 - 53p + 196 = 0$ (b) Solving, $p = 4$ or 49, but $p < 10$, so $p = 4$ $\therefore \quad \det B = 14 - p = 10$ But $\det A = \det B$ So $\quad 2 - 4q = 10 \Rightarrow q = -2$	*In part (a) it is tempting to work with the determinants of the two matrices. But since we are only interested in p, it is worth using the fact that the determinants are equal, and so we only need to work with B.*

1.13 Solving Equations Using Matrices

$\begin{cases} 2x - y = 5 \\ 3x + 4y = 9 \end{cases}$

Solving 2 × 2 equations: Consider the pair of simultaneous equations shown in the notes box. We can rewrite them in a matrix form like this: $\begin{pmatrix} 2 & -1 \\ 3 & 4 \end{pmatrix}\begin{pmatrix} x \\ y \end{pmatrix} = \begin{pmatrix} 5 \\ 9 \end{pmatrix}$. Using matrix algebra, the form of the equation is $MX = A$, where X is the matrix we want to find. Matrix division is not possible, but we can remove the M from the LHS by multiplying by its inverse.

Note that we multiply *in front* on both sides – you could not pre-multiply on the LHS then post-multiply on the RHS.

$$MX = A$$

$$M^{-1}MX = M^{-1}A$$

$$X = M^{-1}A$$

So the solution is found by pre-multiplying A by M^{-1}.

$$\begin{pmatrix} x \\ y \end{pmatrix} = \frac{1}{11}\begin{pmatrix} 4 & 1 \\ -3 & 2 \end{pmatrix}\begin{pmatrix} 5 \\ 9 \end{pmatrix} = \frac{1}{11}\begin{pmatrix} 29 \\ 3 \end{pmatrix}$$

Solution: $x = \frac{29}{11}$ and $y = \frac{3}{11}$ and, as with any sort of equation, you can check your answers by resubstituting the values in the original equation. The equation can also be solved on the calculator: first, enter $A = \begin{pmatrix} 2 & -1 \\ 3 & 4 \end{pmatrix}$ $B = \begin{pmatrix} 5 \\ 9 \end{pmatrix}$, then find the solution by calculating $A^{-1}B$; or by using the simultaneous equation solver. But you need to be able to use all of these methods.

Solving n × n equations: You will always be able to use your GDC to solve simultaneous equations in n variables for $n > 2$: in practice, it is unlikely you will come across anything other than 3 variable equations. The method is the same: $X = M^{-1}A$ or, if the question permits, use the simultaneous equation solver. Try both methods to solve the following system of equations:

$$\begin{cases} 2x + 4y - z = 12 \\ x - y + 4z = 6 \\ 4x + 5y - z = 17 \end{cases}$$

Using the matrix method, $M = \begin{pmatrix} 2 & 4 & -1 \\ 1 & -1 & 4 \\ 4 & 5 & -1 \end{pmatrix}$ and $A = \begin{pmatrix} 12 \\ 6 \\ 17 \end{pmatrix}$.

Then $X = M^{-1}A$.

The solution is $x = 1$, $y = 3$, $z = 2$

Practical application – coding and decoding: Matrix multiplication can be used to code a message which can then be decoded by multiplying by the inverse matrix. It follows that the coding matrix must be square.

Let's use the method to code the following message: "Higher Mathematics is easy". Including the spaces, there are 26 characters in the message. This is converted into a matrix with 3 columns using the numeric equivalents below.

_	A	B	C	D	E	F	G	H	I	J	K	L	M
0	1	2	3	4	5	6	7	8	9	10	11	12	13
N	O	P	Q	R	S	T	U	V	W	X	Y	Z	.
14	15	16	17	18	19	20	21	22	23	24	25	26	27

$$\begin{pmatrix} 8 & 20 & 0 \\ 9 & 8 & 9 \\ 7 & 5 & 19 \\ 8 & 13 & 0 \\ 5 & 1 & 5 \\ 18 & 20 & 1 \\ 0 & 9 & 19 \\ 13 & 3 & 25 \\ 1 & 19 & 0 \end{pmatrix}$$

The message reads down and across, and there is an extra "dummy" space at the end. I have chosen 3 columns so that I can multiply it by a 3 × 3 coding matrix. Any matrix will do – I have chosen one completely at random (but it must have an inverse – some don't).

$$\begin{pmatrix} 8 & 20 & 0 \\ 9 & 8 & 9 \\ 7 & 5 & 19 \\ 8 & 13 & 0 \\ 5 & 1 & 5 \\ 18 & 20 & 1 \\ 0 & 9 & 19 \\ 13 & 3 & 25 \\ 1 & 19 & 0 \end{pmatrix} \begin{pmatrix} 1 & 0 & 3 \\ 2 & -1 & 1 \\ 0 & 2 & 1 \end{pmatrix} = \begin{pmatrix} 48 & -20 & 44 \\ 25 & 10 & 44 \\ 17 & 33 & 45 \\ 34 & -13 & 37 \\ 7 & 9 & 21 \\ 58 & -18 & 75 \\ 18 & 29 & 28 \\ 19 & 47 & 67 \\ 39 & -19 & 22 \end{pmatrix}$$

The matrix on the right represents the encoded message – you will notice that letters which are the same in the original matrix, for example the two 9s in the second row, have been encoded as different numbers.

All that is needed to decode the message is to post-multiply by the inverse of the encoding matrix. Why?

Now, if M is the message matrix and E the encoding matrix, then the coded message is ME. Multiplying by the inverse matrix gives $MEE^{-1} = M$, the original message.

Try decoding $\begin{pmatrix} 19 & 100 \\ -10 & 50 \\ -2 & 76 \\ 7 & 58 \\ -27 & 54 \end{pmatrix}$ if the encoding matrix was $\begin{pmatrix} 1 & 4 \\ -1 & 2 \end{pmatrix}$.

1.14 Eigenvalues and Eigenvectors

Matrices can be used to generate transformations – this is covered in more detail in the chapter on Geometry and Trigonometry. At this point, though, let's look at the eigenvectors and corresponding eigenvalues for a particular 2 × 2 matrix. Consider the matrix $A = \begin{pmatrix} 4 & 1 \\ 3 & 2 \end{pmatrix}$. We want to find any vectors whose direction is unchanged when

Matrix transformations: see page 82

transformed by A. In other words, find x, y, and λ such that $A\begin{pmatrix} x \\ y \end{pmatrix} = \lambda\begin{pmatrix} x \\ y \end{pmatrix}$. This leads to the equation $\begin{pmatrix} 4 & 1 \\ 3 & 2 \end{pmatrix}\begin{pmatrix} x \\ y \end{pmatrix} = \lambda\begin{pmatrix} x \\ y \end{pmatrix}$ and hence:

$$\begin{pmatrix} 4 & 1 \\ 3 & 2 \end{pmatrix}\begin{pmatrix} x \\ y \end{pmatrix} = \begin{pmatrix} 1 & 0 \\ 0 & 1 \end{pmatrix}\begin{pmatrix} x \\ y \end{pmatrix}$$

$$\left(\begin{pmatrix} 4 & 1 \\ 3 & 2 \end{pmatrix} - \lambda\begin{pmatrix} 1 & 0 \\ 0 & 1 \end{pmatrix}\right)\begin{pmatrix} x \\ y \end{pmatrix} = \begin{pmatrix} 0 & 0 \\ 0 & 0 \end{pmatrix}$$

$$\begin{pmatrix} 4-\lambda & 1 \\ 3 & 2-\lambda \end{pmatrix}\begin{pmatrix} x \\ y \end{pmatrix} = \begin{pmatrix} 0 & 0 \\ 0 & 0 \end{pmatrix}$$

> In general, the eigenvalues of A can be found by solving $\det(A - \lambda I) = 0$. This in turn leads to the *characteristic equation* discussed in the next section.

For an infinite number of solutions, $\det\begin{pmatrix} 4-\lambda & 1 \\ 3 & 2-\lambda \end{pmatrix} = 0$ and so $(4-\lambda)(2-\lambda) - 3 = 0$.

Rearranging and simplifying: $\lambda^2 - 6\lambda + 5 = 0$

$$\lambda = 1 \text{ or } \lambda = 5$$

> In other words, any of these vectors will be transformed to multiples of themselves under A.

These are the eigenvalues of A. Substituting into the equation $A\begin{pmatrix} x \\ y \end{pmatrix} = \lambda\begin{pmatrix} x \\ y \end{pmatrix}$ we find that when $\lambda = 1$, $y = -3x$, giving us eigenvectors such as $\begin{pmatrix} 1 \\ -3 \end{pmatrix}$ and $\begin{pmatrix} -2 \\ 6 \end{pmatrix}$ And when $\lambda = 5$, $y = x$ giving us eigenvectors such as $\begin{pmatrix} 1 \\ 1 \end{pmatrix}$ and $\begin{pmatrix} 3 \\ 3 \end{pmatrix}$. It also follows that, if you consider these as position vectors of points on the lines $y = -3x$ and $y = x$, the points will be transformed to new points **on the same lines**. In other words, the lines are invariant when transformed by A. Because $\begin{pmatrix} 0 \\ 0 \end{pmatrix}$ will remain unchanged when multiplied by any matrix, all such lines must pass through the origin.

The line $y = 4x$ is invariant under the transformation with matrix $\begin{pmatrix} 3 & -1 \\ a & -3 \end{pmatrix}$.

(a) Find the value of a.

(b) Find the eigenvalues of the matrix.

(c) Hence find the equation of the second invariant line

(a) $\begin{pmatrix} 3 & -1 \\ a & -3 \end{pmatrix}\begin{pmatrix} x \\ 4x \end{pmatrix} = \lambda\begin{pmatrix} x \\ 4x \end{pmatrix}$

$\begin{cases} 3x - 4x = \lambda x & (1) \\ ax - 12x = 4\lambda x & (2) \end{cases}$

Equation (1) gives $\lambda = -1$

Equation (2) becomes $ax - 12x = -4x \Rightarrow a = 8$

(b) Solve $\det\left(\begin{pmatrix} 3 & -1 \\ 8 & -3 \end{pmatrix} - \begin{pmatrix} \lambda & 0 \\ 0 & \lambda \end{pmatrix}\right) = 0$

$\det\begin{pmatrix} 3-\lambda & -1 \\ 8 & -3-\lambda \end{pmatrix} = 0$

$(3-\lambda)(-3-\lambda) + 8 = 0$

$\lambda^2 - 9 + 8 = 0$

$\lambda = \pm 1$

(c) When $\lambda = 1$, $y = 2x$

In part (a) I have replaced y with 4x in the position vector since we already know the equation of the invariant line.

In part (b) we can make the working quicker by using the "characteristic equation" – see next section.

Characteristic equation of a matrix: If we work out the eigenvalues of the general matrix $A = \begin{pmatrix} a & b \\ c & d \end{pmatrix}$ by solving $\begin{pmatrix} a & b \\ c & d \end{pmatrix}\begin{pmatrix} x \\ y \end{pmatrix} = \lambda \begin{pmatrix} x \\ y \end{pmatrix}$ for λ, we find that $\lambda^2 - (a + d)\lambda + (ad - bc) = 0$; this is known as the *characteristic equation* of the matrix, with the LHS being the *characteristic polynomial*. This gives us a quick way of finding eigenvalues. For example, take the matrix $\begin{pmatrix} 3 & -1 \\ 8 & -3 \end{pmatrix}$ from the question above. When we substitute into the characteristic equation we get $\lambda^2 - (3 + (-3))\lambda + (3 \times (-3) - (-1) \times 8) = 0$ which simplifies to $\lambda^2 - 0\lambda - 1 = 0$, and hence $\lambda = \pm 1$.

The characteristic equation can also be found as $\left| \begin{pmatrix} a - \lambda & b \\ c & d - \lambda \end{pmatrix} \right| = 0$. We can then find an eigenvector for each eigenvalue by solving $\begin{pmatrix} a - \lambda & b \\ c & d - \lambda \end{pmatrix}\begin{pmatrix} x \\ y \end{pmatrix} = \begin{pmatrix} 0 \\ 0 \end{pmatrix}$.

Powers of matrices: Matrices have a wide range of applications, and it is often the case that the powers of matrices can be useful, for example in networks and in probability. Eigenvalues and eigenvectors can be used to simplify the process of finding any power of a matrix.

Let's continue with our study of the matrix $A = \begin{pmatrix} 4 & 1 \\ 3 & 2 \end{pmatrix}$. Look back to the start of this section and you will see that A has eigenvalues 1 and 5 with corresponding eigenvectors $\begin{pmatrix} 1 \\ -3 \end{pmatrix}$ and $\begin{pmatrix} 1 \\ 1 \end{pmatrix}$, and follow these steps to find A^n.

Step 1: Create the *diagonal matrix* $D = \begin{pmatrix} 1 & 0 \\ 0 & 5 \end{pmatrix}$. The two numbers on the leading diagonal are the eigenvalues of A, the other elements are set to 0.

Step 2: Create the columns of matrix P from the eigenvectors.

$$P = \begin{pmatrix} 1 & 1 \\ -3 & 1 \end{pmatrix}.$$

Step 3: It will always be true that $AP = PD$ (this follows from the definitions of eigenvalues and eigenvectors) and a little bit of matrix algebra takes us to the general result:

$A^n = PD^nP^{-1}$ *(this is in the formula book)*

> The eigenvector in the first column of P corresponds to the eigenvalue in the first column of D. There are of course any number of eigenvectors for a particular eigenvalue – choose the easiest!

And since D is always of the form $\begin{pmatrix} \lambda_1 & 0 \\ 0 & \lambda_2 \end{pmatrix}$, $D^n = \begin{pmatrix} \lambda_1^n & 0 \\ 0 & \lambda_2^n \end{pmatrix}$. So let's use step 3 to find A^3. First we need to write down the inverse of P.

$\det(P) = 4$, so $P^{-1} = \frac{1}{4}\begin{pmatrix} 1 & -1 \\ 3 & 1 \end{pmatrix}$.

Then:

$$\begin{pmatrix} 4 & 1 \\ 3 & 2 \end{pmatrix}^3 = \begin{pmatrix} 1 & 1 \\ -3 & 1 \end{pmatrix}\begin{pmatrix} 1^3 & 0 \\ 0 & 5^3 \end{pmatrix} \times \frac{1}{4}\begin{pmatrix} 1 & -1 \\ 3 & 1 \end{pmatrix}$$

$$= \frac{1}{4}\begin{pmatrix} 1 & 125 \\ -3 & 125 \end{pmatrix}\begin{pmatrix} 1 & -1 \\ 3 & 1 \end{pmatrix}$$

$$= \begin{pmatrix} 94 & 31 \\ 93 & 32 \end{pmatrix}$$

Although you can of course use your GDC to calculate powers of matrices, a question may ask you to investigate a long term trend by looking at a matrix to the power n, in which case you can use the diagonal matrix method above; or you may be asked to find the nth power of a matrix.

Practical application – dynamic system: Consider a reversible chemical reaction between two substances A and B, in which a proportion of A turns into B and a proportion of B turns into A. A *transition matrix* neatly shows this information: matrix T shows that in the reaction, 80% of A turns into B, whereas 60% of B turns into A. The remainder of the substances stay unchanged.

$$\text{To} \begin{array}{c} \\ A \\ B \end{array} \begin{pmatrix} 0.2 & 0.4 \\ 0.8 & 0.6 \end{pmatrix} = T$$

with From A B above.

Let us suppose that we start with a mixture of 50% A and 50% B – we can find the proportions of each substance after the reaction by using matrix multiplication.

$$\begin{pmatrix} 0.2 & 0.4 \\ 0.8 & 0.6 \end{pmatrix}\begin{pmatrix} 0.5 \\ 0.5 \end{pmatrix} = \begin{pmatrix} 0.3 \\ 0.7 \end{pmatrix}$$

Thus the mixture now contains 30% of A and 70% of B.

Multiplying by the transition matrix again is equivalent to running the reaction a second time:

$$\begin{pmatrix} 0.2 & 0.4 \\ 0.8 & 0.6 \end{pmatrix}\begin{pmatrix} 0.3 \\ 0.7 \end{pmatrix} = \begin{pmatrix} 0.34 \\ 0.66 \end{pmatrix}$$

We could also have run the reaction twice by multiplying by T^2. That is, the proportion of A and B after two reactions is:

$$\begin{pmatrix} 0.2 & 0.4 \\ 0.8 & 0.6 \end{pmatrix}\begin{pmatrix} 0.2 & 0.4 \\ 0.8 & 0.6 \end{pmatrix}\begin{pmatrix} 0.5 \\ 0.5 \end{pmatrix} = \begin{pmatrix} 0.36 & 0.32 \\ 0.64 & 0.68 \end{pmatrix}\begin{pmatrix} 0.5 \\ 0.5 \end{pmatrix} = \begin{pmatrix} 0.34 \\ 0.66 \end{pmatrix}.$$

What happens if we run the reaction n times? We need to find T^n. Here are the steps:

1. Write down the characteristic equation.
2. Solve to find the eigenvalues.
3. Hence write down an eigenvector for each eigenvalue.
4. Write down the diagonal matrix D, and hence D^n.
5. Write down the matrix of eigenvectors P.
6. Work out P^{-1}.
7. Write down PD^nP^{-1} and simplify.

Solution:

1. $\lambda^2 - (0.2 + 0.6)\lambda + (0.12 - 0.32) = 0$

   $\lambda^2 - 0.8\lambda - 0.2 = 0$

2. Eigenvalues are 1 and –0.2.

3. $\lambda = 1$ leads to $y = 2x$, hence an eigenvector is $\begin{pmatrix} 1 \\ 2 \end{pmatrix}$

   $\lambda = -0.2$ leads to $y = -x$, hence an eigenvector is $\begin{pmatrix} 1 \\ -1 \end{pmatrix}$

4. $D = \begin{pmatrix} 1 & 0 \\ 0 & -0.2 \end{pmatrix}$, $D^n = \begin{pmatrix} 1 & 0 \\ 0 & (-0.2)^n \end{pmatrix}$

5. $P = \begin{pmatrix} 1 & 1 \\ 2 & -1 \end{pmatrix}$

6. $P^{-1} = -\frac{1}{3}\begin{pmatrix} -1 & -1 \\ -2 & 1 \end{pmatrix}$

7. $$PD^nP^{-1} = -\frac{1}{3}\begin{pmatrix} 1 & 1 \\ 2 & -1 \end{pmatrix}\begin{pmatrix} 1 & 0 \\ 0 & (-0.2)^n \end{pmatrix}\begin{pmatrix} -1 & -1 \\ -2 & 1 \end{pmatrix}$$

$$= -\frac{1}{3}\begin{pmatrix} 1 & (-0.2)^n \\ 2 & -(-0.2)^n \end{pmatrix}\begin{pmatrix} -1 & -1 \\ -2 & 1 \end{pmatrix}$$

$$= -\frac{1}{3}\begin{pmatrix} -1-2(-0.2)^n & -1+(-0.2)^n \\ -2+2(-0.2)^n & -2-(-0.2)^n \end{pmatrix}$$

You might like to confirm that putting $n = 2$ results in the matrix T^2 we found above.

What happens as $n \to \infty$? $(-0.2)^n \to 0$, so $T^n \to \begin{pmatrix} \frac{1}{3} & \frac{1}{3} \\ \frac{2}{3} & \frac{2}{3} \end{pmatrix}$ which creates a steady state: $\frac{1}{3}$ of both A and B remain the same, the other $\frac{2}{3}$ swapping over. So, after many reactions, $\begin{pmatrix} \frac{1}{3} & \frac{1}{3} \\ \frac{2}{3} & \frac{2}{3} \end{pmatrix}\begin{pmatrix} 0.5 \\ 0.5 \end{pmatrix} = \begin{pmatrix} \frac{1}{3} \\ \frac{2}{3} \end{pmatrix}$, which means that the mixture now contains A:B in the ratio 1:2, and this ratio remains constant every time the reaction takes place.

Eigenvalues and Eigenvectors: Practice Exercise

1. For the matrix $\begin{pmatrix} 5 & 2 \\ 3 & 4 \end{pmatrix}$

   (a) Find the eigenvalues

   (b) Find the eigenvectors for each eigenvalue, expressed in a general form.

2. The line $y = 3x$ is invariant under the transformation with matrix $\begin{pmatrix} 4 & -1 \\ -3 & a \end{pmatrix}$. Find the value of a.

3. $A = \begin{pmatrix} 3 & -2 \\ 3 & -4 \end{pmatrix}$. Find vectors P and D such that $A^n = PD^nP^{-1}$, and hence evaluate A^4.

Answers

1. (a) 2, 7. (b) $\begin{pmatrix} x \\ -\frac{3}{2}x \end{pmatrix}$, $\begin{pmatrix} x \\ x \end{pmatrix}$

2. $a = 2$

3. $D = \begin{pmatrix} -3 & 0 \\ 0 & 2 \end{pmatrix}$;

   $P = \begin{pmatrix} 1 & 3 \\ 2 & 1 \end{pmatrix}$

   Other answers are possible

   $A^4 = \begin{pmatrix} 3 & 26 \\ -39 & 94 \end{pmatrix}$

Number and Algebra: Long Answer Questions

There follows a selection of Paper 2 style exam questions related to the Number and Algebra topic, although some may require knowledge and techniques from other areas of the syllabus.

The answers are given here, but full working may be found on the Peak Study Resources website (www.peakib.com).

1. Four friends are discussing how their parents have paid them a monthly allowance over the previous two years.

    Alan says that he has received a fixed amount of $55 each month.

    (a) Calculate the total Alan has received over the two years.

    Bryony received $8 in the first month, $12 in the second and then an extra $4 every month for the two years.

    (b) Calculate:

      (i)   how much Bryony received in the 13th month;

      (ii) the total she received over the two years.

    Carl received $10 in the first month, increasing by 12% every month.

    (c) Calculate, to the nearest dollar:

      (i)   how much Carl received in the 8th month;

      (ii) the total he received over the two years.

    Drishti received $1100 and then invested it at a fixed rate of interest, ***compounded monthly*** for two years. She calculates that the total she received was the same as the mean total received by the other three.

    (d) Calculate:

      (i)   how much Drishti received in total, to the nearest dollar;

      (ii) the annual interest rate for her investment, correct to 3SF.

    _____

    Answers:

      (a) $1320

      (b) (i) $56   (ii) $1296

      (c) (i) $22   (ii) $1182

      (d) (i) $1266 (ii) 7.04%

2. **Part A**

The function $f(x) = ax^3 + bx - 3$. The points $(1, -6)$ and $(2, 15)$ lie on the graph of f.

(a) Form two equations for a and b.

(b) Find the values of a and b.

(c) Hence find the points where the graph of f intersects the x-axis.

Part B

Consider the arithmetic sequence 3, 8, 13, …

(a) Find:

    (i) the value of the 15th term

    (ii) the sum of the first 15 terms

Chloe calculates that the formula for the sum of the first n terms is $\frac{n}{2}(5n + 1)$.

(b) Show that Chloe's formula works for $n = 3$.

(c) For another value of n the sum of the series is 1010.

    (i) Use Chloe's formula to show that $5n^2 + n - 2020 = 0$

    (ii) Hence find the value of n.

_____

Answers:

Part A

(a) $-6 = a + b - 3$ or $a + b = -3$

    $15 = 8a + 2b - 3$ or $8a + 2b = 18$

(b) $a = 4$, $b = -7$

(c) $(-1, 0)$, $(-\frac{1}{2}, 0)$, $(1\frac{1}{2}, 0)$

Part B

(a) (i) 73   (ii) 570

(c) (ii) $n = 20$

3. (a) Write $30 + 40\,e^{\frac{\pi}{3}i}$ in the form $re^{i\theta}$.

    Two AC voltage sources are connected in a circuit. Their equations are given as:

    $$V_1 = 30\cos(60t) \text{ and } V_2 = 40\cos\left(60t + \frac{\pi}{3}\right)$$

    (b) State the period of the voltage oscillations.

    (c) Find an expression for the total voltage (V) in the circuit in the form $V = A\cos(Bt + C)$

    (d) Hence state the maximum voltage in the circuit.

    The average power in a circuit over a time period T is given by $P_{av} = \frac{1}{RT}\int_0^T (v(t))^2 \, dt$ where R represents the fixed resistance in the circuit.

    (e) Given there is a fixed resistance of 2Ω in the circuit described above, find the average power, in W, in the circuit over one period. Give your answer to 3SF.

    _____

    Answers:

    (a) $60.8276\,e^{0.60589i}$ (GDC)

    (b) $\frac{2\pi}{60} \approx 0.1047$

    (c) $V = 60.8276\cos(60t + 0.60589)$

    (d) $60.827\,\text{V}$

    (e) $925\,\text{W}$

4. (a) Find the eigenvalues and corresponding eigenvectors for the matrix $\begin{pmatrix} 1.1 & 0.12 \\ 0 & 1.08 \end{pmatrix}$.

A farmer breeds sheep on two farms, A and B. The number of sheep on farm A increase by 10% each year and the number of sheep on farm B by 20%. To maintain a balance in the number of sheep on each farm at the end of each year 10% of the sheep from farm B are moved to farm A.

After n years let the number of sheep in farm A be a_n and the number of sheep in farm B be b_n. At the start of the first year there are a_0 and b_0 sheep on farms A and B respectively.

(b) Show that $\begin{pmatrix} a_1 \\ b_1 \end{pmatrix} = \begin{pmatrix} 1.1 & 0.12 \\ 0 & 1.08 \end{pmatrix} \begin{pmatrix} a_0 \\ b_0 \end{pmatrix}$

(c) By writing $\begin{pmatrix} 1.1 & 0.12 \\ 0 & 1.08 \end{pmatrix}$ as a product of 3 matrices find an expression for $\begin{pmatrix} 1.1 & 0.12 \\ 0 & 1.08 \end{pmatrix}^n$

Given $a_0 = 50$ and $b_0 = 200$

(d) (i) Find an expression for a_n and b_n

(ii) After how many years does the number of sheep on farm A exceed the number of sheep on farm B?

(e) The maximum number of sheep that can be supported by farm A is 1000. For how many years can the farmer use the policy described in this question?

Answers:

(a) $\lambda = 1.1$ and 1.08 with corresponding eigenvectors $\begin{pmatrix} 1 \\ 0 \end{pmatrix}$ and $\begin{pmatrix} -6 \\ 1 \end{pmatrix}$

(c) $\begin{pmatrix} 1.1 & 0.12 \\ 0 & 1.08 \end{pmatrix}^n = \begin{pmatrix} 1.1^n & 6(1.1^n - 1.08^n) \\ 0 & 1.08^n \end{pmatrix}$

(d) (i) $a_n = 50 \times 1.1^n + 1200(1.1^n - 1.08^n)$

$b_n = 200 \times 1.08^n$

(ii) 7 years

(e) 12 years

Chapter 2: FUNCTIONS

2.1 Basics of Functions

A *relation* is an algebraic rule which shows how one set of numbers is related to, or obtained from another set. Relations often model real-life situations, so it is necessary to understand the notation used and the different types of relation which occur and the notation used.

One-to-one and many-to-one: When two sets of numbers are related, their relationship can be defined in one of four ways:

- *One-to-one*: Each object has one image and vice versa.

  Examples:
  $x \rightarrow x + 1$
  Name of person $\rightarrow$ passport number

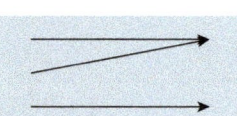

- *Many-to-one*: Each object has only one image but different objects can map onto the same image.

  Examples:
  $x \rightarrow$ nearest integer to x
  Student $\rightarrow$ mark in mathematics exam

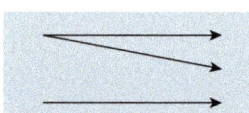

- *One-to-many*: An object can have more than one image but each image is related to only one object.

  Examples:
  $x \rightarrow \pm\sqrt{x}$
  Father $\rightarrow$ children

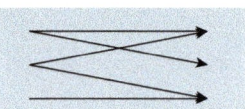

- *Many-to-many*: Each object can be related to several images and each image can be related to several objects.

  Examples:
  $x \rightarrow$ prime factors of x
  Vegetable $\rightarrow$ possible colours of vegetable

Functions are defined as relationships which are either one-to-one or many-to-one.

Function notation: A function is defined using the notation $f(x)$, but note that other letters may be used, particularly if modelling physical quantities. For example, the velocity of an object at a particular time may be defined using $v(t)$; the cost of buying a number of articles could be given by $C(n)$.

Thus the function $f(x) = 3x^2 - 1$ can be read as: "Function f takes any number, x, and turns it into $3x^2 - 1$." The function notation is also used with specific numbers; for example, $f(2) = 3 \times 2^2 - 1 = 11$.

Domain: The set of values to be input to a function is called the *domain.* In many functions, *any* value can be input, in which case the domain is $x \in \mathbb{R}$. However, the domain may be restricted for two reasons:

- Certain values of x may give impossible results, such as division by 0 or the square root of a negative number. For example, x cannot take the value 4 in the function $f(x) = \dfrac{x}{x-4}$, and this would be written as: $f(x) = \dfrac{x}{x-4}$ for $x \neq 4$.
- For the purposes of a particular question the domain may be "artificially" restricted. If $g(x) = 2x^2 - 3$ for $x > 0$, the function would only take positive values of x.

It's important to note the domain in an exam question because it may affect your answer. Solving $g(x) = 5$ in the example above would lead to the solution $x^2 = 4$ and hence $x = 2$; $x = -2$ is not in the domain.

Range: The set of values produced by a function is called the range. In the examples above, the range of g is $g(x) > -3$ since there is a minimum at $(0, -3)$. The range of f is best found by inspecting a graph: the range is the complete set of possible y values, so in this case would be $f(x) \neq 1$.

> If you draw the graph of f you will see there is a horizontal asymptote at $y = 1$. All other y values are represented on the graph.

The next sections use the image of a "function machine" which represents a function as a black box.

When the handle is turned, the 5 drops in the top and the function machine turns it into an 11!

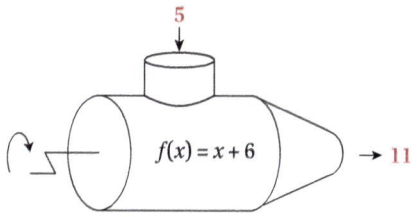

Inverse functions: An inverse function "reverses" the effect of a function. The inverse of add 2 is subtract 2. The inverse of squaring is square rooting. In terms of the function machine, just turn the handle the other way and the 11 turns back into a 5. The notation for an inverse function is $f^{-1}(x)$.

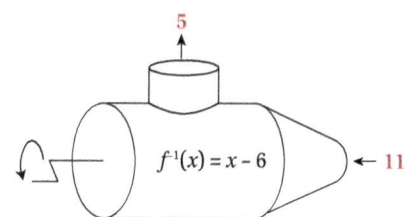

To work out the inverse of a function – particularly a more complex one – the method is:

- Write the function in the form $y = $ the function
- Replace the y with an x and all the x's with y's.
- Make y the subject – you will have the inverse function.
- Write down the inverse function starting with $f^{-1}(x) = $ A few points about inverse functions which you need to know:
- The graph of an inverse function can be found by reflecting the graph of the function in the line $y = x$.
- The statement $f(5) = 11$ is exactly the same as $f^{-1}(11) = 5$.
- The domain of a function is the same as the range of its inverse, and the range of a function is the same as the domain of its inverse.
- $(f \circ f^{-1})(x) = (f^{-1} \circ f)(x) = x$ (eg: $\sqrt{x^2} = \left(\sqrt{x}\right)^2 = x$)

Example: Given that $f^{-1}(-8) = 3$, find the value of a in the function $f(x) = 10 - ax^2$.

Solution: There is no need to find the inverse function.

If $f^{-1}(-8) = 3$ then it follows that $f(3) = -8$.

Thus, $10 - a \times 3^2 = -8$ giving $a = 2$.

Given the function $f(x) = \sqrt{x+2}$ for $x \geq -2$,

    (a)   Find the inverse function $f^{-1}(x)$

    (b)   Find the domain of f^{-1}

(a) $y = \sqrt{x+2}$
 	
     $x = \sqrt{y+2}$
 	
     $x^2 = y + 2$
 	
     $y = x^2 - 2$
 	
     So $f^{-1}(x) = x^2 - 2$
 	
(b) Domain of f^{-1} is $x \geq 0$
	(a) Use the steps above to find the inverse. Note that the algebra is sometimes more complicated. For example, try finding the inverse of $f(x) = \dfrac{2x+4}{x-1}$.
	
	(b) Domain of inverse = range of function.

See the website for full working of the inverse example in my note.

The inverse of a many-to-one function: The inverse of a one-to-one function is itself one-to-one and therefore also a function. However, the inverse of a many-to-one is a one-to-many and hence ***not*** a function, unless the domain of the original is restricted, hence turning it into a one-to-one function.

For example, the graph shows a quadratic with a vertex at $x = -2$.

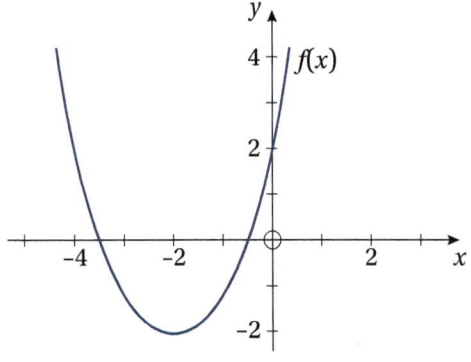

If a function has turning points it is many-one. To ensure the inverse is a function, the domain should be restricted so as not to include any turning points.

There are no domain restrictions. However, if the inverse is to be a function, we must restrict the domain of the quadratic to remove the turning point. So, a suitable restriction would be $x \geq -2$. When reflected in the line $y = x$ (to get the graph of the inverse function) the result is:

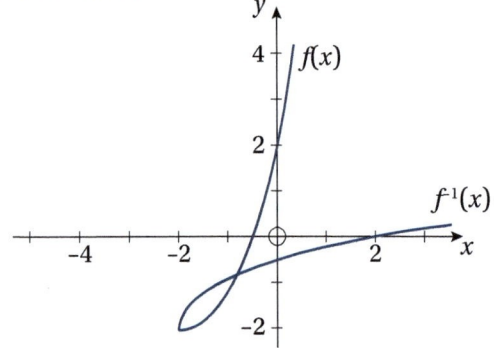

The inverse is a function since it is a one-to-one relationship.

Composite functions: If the values of one function are input to another one, the result is a *composite function*. Given $f(x) = x^2$ and $g(x) = x - 1$ then $f(g(3)) = f(2) = 4$. This is not the same as $g(f(3)) = g(9) = 8$. It is important to understand that the functions are not being multiplied together; a number is being put through one function, then the other. This can be illustrated using the function machines again.

Remember that:
$(f^{-1} \circ f)(x) = (f \circ f^{-1})(x) = x$

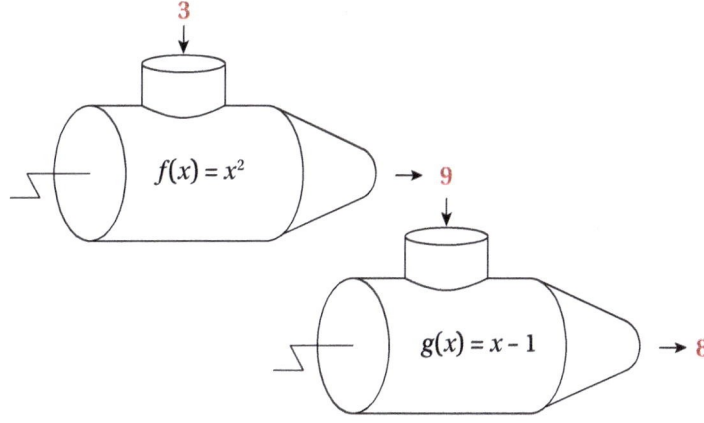

The actual notation used (to avoid multiple brackets) is $(g \circ f)(3)$. Say this as "g of f of 3" and remember that 3 is put into f first and then into g.

To find $(g \circ f)(x)$, work like this: $(g \circ f)(x) = g(f(x)) = g(x^2)$. Now, in words, function g is "subtract 1." So $g(x^2) = x^2 - 1$. Try using the same method to show that $(f \circ g)(x) = (x - 1)^2$.

Questions may also involve composites such as $f \circ f$ (put the value through function f twice), or $f^{-1} \circ g$ (put the value through function g then through the inverse of function f). Try the following examples:

$f(x) = 3x^2$, $g(x) = \frac{1}{x}$, find $(g \circ f)(x)$ $\frac{1}{3x^2}$

$f(x) = x^2$, $g(x) = \sin x$, find $(f \circ g)\left(\frac{2\pi}{3}\right)$ $\frac{3}{4}$

$f(x) = x + 4$, $g(x) = e^x$, find $(g^{-1} \circ f)(x)$ $\ln(x + 4)$

$f(x) = x + 1$, find $(f \circ f)(x)$ $x + 2$

$f(x) = x + 3$, $g(x) = 2^x$, solve $(f \circ g)(x) = (g \circ f)(x)$ -1.22 (GDC)

> Given that $f(x) = 4(x - 1)$ and $g(x) = \dfrac{6 - x}{2}$
>
>   (a)   Find g^{-1}
>
>   (b)   Solve $(f \circ g^{-1})(x) = 4$

(a) $y = \dfrac{6 - x}{2}$ $\quad x = \dfrac{6 - y}{2}$ $\quad 2x = 6 - y$ $\quad y = 6 - 2x$ So $g^{-1}(x) = 6 - 2x$ (b) $f(g^{-1}(x)) = 4$ $\quad f(6 - 2x) = 4$ $4(6 - 2x - 1) = 4$ $\quad 6 - 2x - 1 = 1$ $\quad\quad 2x = 4$ $\quad\quad x = 2$	*(b) Just to reiterate – do not multiply the functions together. Put one function as the input to the second.*

Practical application of composite functions: A composite function can be used to model a real world situation where "something depends on something else which depends on something else!" For example, suppose Maria works in a technology shop earning a basic salary of $450 per week, and also 3% commission on her total sales over $4000 per week.

  (a)   How much did she earn in total when her sales for the week were $6250?

       Answer: $450 + 0.03 \times 2250 = \$517.50$.

  (b)   The composite function $E = p + q(x - r)$ represents her total earnings in a week where she had $x of sales. Assuming $x > 4000$, state the values of p, q and r.

       Answer: $p = 450$, $q = 0.03$, $r = 4000$

  (c)   Using your values of p, q and r, express x in terms of E.

       Answer: $x = \dfrac{E - 450}{0.03} + 4000$

  (d)   Hence find her sales in a week where her earnings were $555.

       Answer: $x = \dfrac{555 - 450}{0.03} + 4000 = \$7500$

Using a graph to answer function questions: A graph is effectively a "picture" of a function. The x-axis contains the numbers input to the function and therefore represents the domain; the y-axis contains the resulting function values, and therefore shows the range. Thus a point with coordinates (a, b) is the graphical equivalent of $f(a) = b$, and also $f^{-1}(b) = a$. For example, the graph of $y = 2^x$ is a "picture" of the function $f(x) = 2^x$, and the point $(3, 8)$ represents $f(3) = 8$. It is helpful to think of the y-axis as the "function axis".

You can answer questions about a function just from its graph, even without an equation. For example, look at this graph. Because there is a point $(5, 3)$, we know that $f(5) = 3$. What about $f^{-1}(-1)$? Work backwards from -1 on the y-axis, and we can see that $f^{-1}(-1) = -3$.

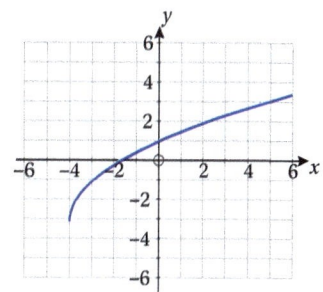

Here are some more possible questions:

Find $(f \circ f)(-4)$.   Answer: $(f \circ f)(-4) = f(-3) = -1$

Solve $f^{-1}(a) = 0$.   Answer: $a = f(0) = 1$

What is the domain of f^{-1}?   Answer: same as the range of f, so $-3 \le x \le 3$

You may also be asked to sketch the graph of the inverse function. The easiest way to do this is to take key points, such as $(5, 3)$, $(2, 2)$, $(0, 1)$, $(-3, -1)$, $(-4, -3)$ reverse their coordinates – $(3, 5)$, $(2, 2)$, $(1, 0)$ etc – plot them and join them up.

2.2 Graphs of Functions

A graph is an excellent tool for interpreting a function. From a graph we can see when the function is increasing or decreasing, what the range of the function is, where it cuts the axes and so on. Therefore it is important to be able to sketch and understand graphs of different types of functions. Remember that your calculator can be of great benefit, and you should fully understand its graphing functions (see page 39); but you must also be able to sketch graphs without a calculator.

Note the difference between "draw" and "sketch". Drawing a graph will require plotting many points. A sketch shows the shape of a graph and how it relates to the axes, with a few key points marked in.

Asymptotes: A graph such as $y = 2^x$ has a horizontal asymptote because as x gets smaller, the values of y get ever closer to 0 without ever reaching it. Some functions have graphs with vertical asymptotes which arise because division by 0 is impossible.

For example, $y = \dfrac{x}{2 - x}$ (pictured) has a vertical asymptote at $x = 2$; as x gets closer to 2, the denominator gets closer to 0. This graph also has a horizontal asymptote at $y = -1$. Try putting x values of 10, 100 and 1000 into the function to see why.

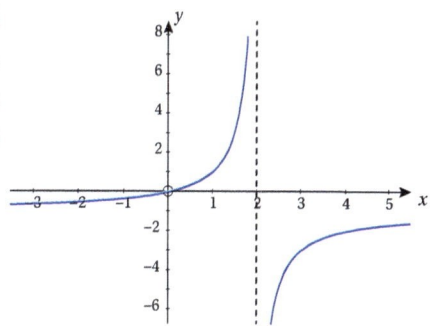

Significant features: The other main features of a graph you should be able to identify are the x-intercepts (where $y = 0$); the y-intercept (where $x = 0$); and turning points, where the gradient $= 0$. But note that a graph doesn't necessarily have all, or even any, of these features.

Transformations of graphs: You should be able to sketch the graphs of the basic functions $y = x^2$, $y = x^3$, $y = \dfrac{1}{x}$, $y = a^x$, $y = \log_{10}x$ and $y = \ln x$. The effect of simple numerical changes to these functions (involving additions, multiplications and minus signs) results in specific, simple transformations, thus extending the number of functions which can be easily sketched.

The graph transformations you need to know are:

Change to function	Transformation
$y = f(x) + a$	Translate (slide) graph upwards by a units
$y = f(x + a)$	Translate graph to the left by a units
$y = af(x)$	Stretch graph vertically by scale factor a
$y = f(ax)$	Stretch graph horizontally by scale factor $\frac{1}{a}$
$y = -f(x)$	Reflect graph in x-axis
$y = f(-x)$	Reflect graph in y-axis

Transformations in the x direction always do the opposite of what you expect!

For example, $y = (x - 1)^2 + 2$ will translate the graph of $y = x^2$ to the right by 1 and up by 2, that is, a translation of $\binom{1}{2}$.

$y = -\dfrac{3}{x + 2}$ is a composite transformation of $y = \frac{1}{x}$. To obtain the correct order of transformations, consider what order you would work out the expression if you put in a value for x. This would be:

- Add 2 to x
- Multiply the function by 3
- Change sign

Be aware of the difference between, say, adding 2 to the x part of the function, and adding 2 to the *whole* function.

The equivalent transformations are:

- Translate left 2 units
- Stretch by 3 in the y direction
- Reflect in the x-axis

If a question asks you to sketch the transformation of a particular curve, the best way is to work out where the key points on the graph will move to, plot them, and then sketch the curve. (Of course, if the question allows it, the best way is to use a GDC!) And sometimes the transformation is not specifically mentioned.

We could also use a GDC for the first part; but because of the "hence" we must consider transformations for the second part.

Example: Find the turning point on the graph of $f(x) = x^2 + \dfrac{16}{x} - 9$ and hence write down the turning point on the graph of $f(x) = x^2 + \dfrac{16}{x}$.

Solution: Using differentiation, we find the turning point is when $x = 2$. Substitute this value into the function, and the turning point is (2, 3). Now to get the second function we add 9 to the first, and so the graph has been translated up 9. The turning point goes with it, and is now at (2, 12).

The graph of $f^{-1}(x)$: Consider the graph of $y = x^2$, $x > 0$ (which represents the function $f(x) = x^2$). When $x = 3$, $y = 9$ (ie $f(3) = 9$). The graph of the inverse function is $y = \sqrt{x}$, and when $x = 9$, $y = 3$. *Any* point (a, b) on the graph of f becomes (b, a) on the graph of f^{-1}. So the graph of $y = f^{-1}(x)$ is always the reflection of the graph of $y = f(x)$ in the line $y = x$.

Transformations of graphs: Practice exercise

1. Find the x-intercepts, y-intercept on the graph of $y = x^2 + x - 6$. Hence find the images of these points when transformed onto the graph of $y = x^2 + x - 3$.

2. The graph of $y = 2\ln x$ is obtained from the graph of $y = \ln(x - 3)$. List the transformations required.

3. The points $(0, 3)$ and $(-2, 4)$ lie on the graph of $f(x)$. Write down the equivalent points on the graphs of: (a) $f(2x)$; (b) $-f(x + 1)$

4. Find the turning point on the graph of $f(x) = -x^2 + 4x - 5$. The graph of f is translated to the graph of g by the vector $\begin{pmatrix} 0 \\ a \end{pmatrix}$. Find the values of a such that $g(x) = 0$ has exactly two solutions.

5. The graph of $f(x) = \frac{1}{x}$ is transformed into the graph of $g(x)$ by a translation along the vector $\begin{pmatrix} -3 \\ 1 \end{pmatrix}$. Find an expression for $g(x)$ and write down the equations of the asymptotes of the graph of g.

Answers

1. $(2, 0), (-3, 0), (0, -6)$;
   $(2, 3), (-3, 3), (0, -3)$.

2. Translation $\begin{pmatrix} -3 \\ 0 \end{pmatrix}$ then stretch ×2 in the y direction.

3. (a) $(0, 3), (-1, 4)$
   (b) $(-1, -3), (-3, -4)$

4. $(2, -1)$, $a > 1$

5. $y = \frac{1}{x + 3} + 1 = \frac{x + 4}{x + 3}$
   $x = -3, y = 1$

Functions and Graphs with a GDC: Entering functions into your GDC, then drawing graphs and solving equations, are all essential techniques which you should be able to carry out with accuracy and fluency.

Function keys: You should be able to enter a wide range of functions with confidence – you don't want to spend time in the exam searching your calculator. Look at the list in the notes box – make sure you can key in each of the functions. You should also be able to work in fraction notation, and be able to enter functions such as $y = \sqrt[3]{\dfrac{x}{x - 1}}$.

$\sqrt[3]{12}$

$\arccos 128°$

e^4

$\ln(2^6)$

Tables: GDCs have the facility to work out a table of values for a function. Having input the function in the form $y = f(x)$ you can set up a table by selecting the first x value and then the steps by which you want x to increase. In this example, the function $y = 2 - 3\sin x$ has been entered into the function editor, and then a table created starting with $x = 0$ and increasing x in steps of 30°. This can be helpful if you need to know several values, if you want to plot a graph by hand or if you're having difficulty creating the appropriate scales for a calculator plot – the table indicates the lowest and highest values of y and helps you set an appropriate window.

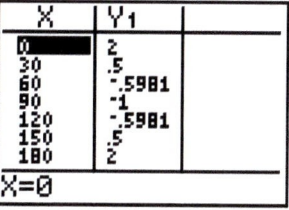

Drawing graphs: Three important points to remember when drawing and using GDC graphs:

- Make sure the function you type into the editor is actually the same as in the question. You may, for example, have to use brackets which aren't actually required on the written page. 2^{x+3}, if typed as 2 ^ x + 3, will calculate as $2^x + 3$. To get the correct answer you would need a bracket: 2 ^ $(x + 3)$.

- The GDC has a few standard sets of scales, but you will probably have to set up the "window" yourself in order to see the required part of the graph. You may well have to zoom into a part of the graph to see exactly what is happening. The two screenshots are of the same graph, but only the lower one shows the intersections with the x-axis.

- The GDC can give you the values of key points such as intersections with the axes, points where lines intersect, turning points and so on. If you want to read off your

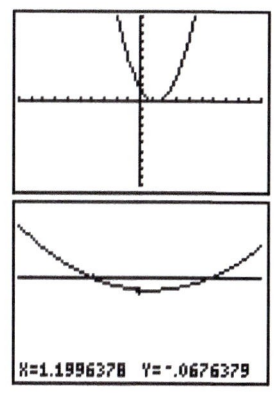

Most modern calculators will allow the entry of expressions in their correct format, for example 2^{x+3} instead of $2^{\wedge}(x + 3)$. It is still important to ensure the correct use of brackets.

own point, make sure you know the scales being used, ie how much each mark on the axes is worth.

I've sketched two graphs and you can see my sketches below, alongside the graphs on the GDC. The two equations are: $y = x^3 - 3x + 1$ and $y = 2^x + \dfrac{1}{x+1}$. In both cases, I used $-2 \le x \le 2$.

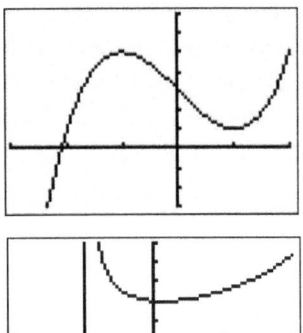
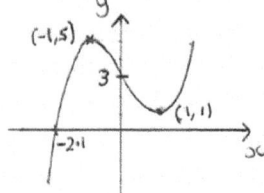
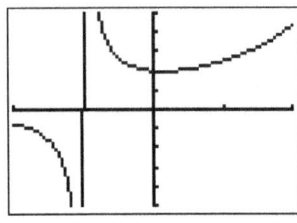
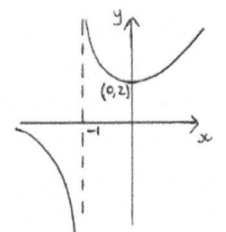

Solving equations: GDCs have built in equation solvers. However, if you already have the graph of the relevant function, then it can be used to solve equations. The easiest way to do this is to ensure your equation has a 0 on the right hand side because then all you have to do is find out where the graph cuts the axis. For example, solve $x^2 - 2 = \dfrac{1}{x}$, $x > 0$.

If you have already entered one of the functions into your GDC, then using graph intersections will be the simplest way to solve an equation.

First we need to rewrite this equation as $x^2 - 2 - \dfrac{1}{x} = 0$ and draw the graph of $y = x^2 - 2 - \dfrac{1}{x}$.

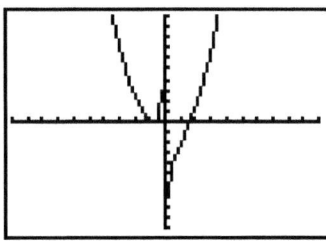

Now use the "zero" or "root" feature to find where the graph cuts the x-axis and this will be the solution to the equation which is $x = 1.618$.

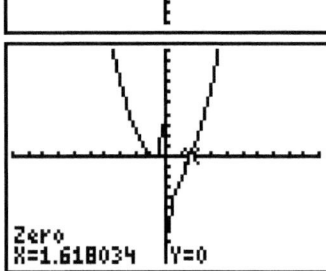

Given that $f(x) = x^3 \times 2^{-x}$, $x \geq 0$

    (a)   Sketch the graph of $f(x)$ showing its asymptotic behaviour.

    (b)   Find the co-ordinates of the maximum point, and hence state the range of $f(x)$.

    (c)   Draw a line on your graph to show that $f(x) = 1$ has two solutions.

    (d)   Find the solutions to $f(x) = 1$ giving your answers to 3 significant figures.

(a) and (c)

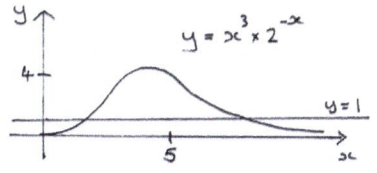

(b)   Maximum = (4.33, 4.04)

       Range is $0 \leq f(x) \leq 4.04$

(c)   The line $y = 1$ intersects $f(x)$ at two points.

(d) $x = 1.37$ or 9.94

(a) When you sketch a graph, you must label the scales and the graph, and show at least one point on each axis to give an indication of scale. Do not go beyond the domain.

(b) and (d) You should be able to use your GDC to find turning points on a graph, and to find where two graphs intersect.

Two functions are given by $f(x) = \dfrac{2}{x^2 + 1}$ and $g(x) = \sqrt{x+1}$.

    (a)   Sketch the graphs of $f(x)$ and $g(x)$ on the same diagram, with values of x between -3 and 3, and values of y between -1 and 3. Label each curve and any axis intercepts.

    (b)   Explain why there cannot be any solutions to $g'(x) = 0$ in the range $-3 \leq x \leq 3$.

    (c)   Find the solution to the equation $\dfrac{2}{x^2 + 1} - \sqrt{x+1} = 0$. Give your answer to 4 significant figures.

(a)

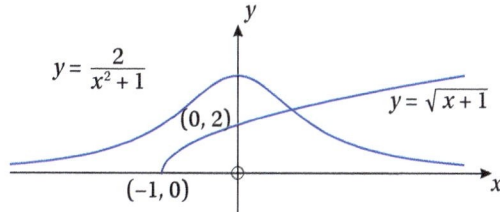

(b)   There are no turning points on the graph of g.

(c) $x = 0.7235$ (GDC)

(b) The relationship between the graph of a function and the derived function is explained in the Calculus chapter.

2.3 Mathematical Modelling

Modelling is at the heart of most applications of mathematics. At its simplest, a mathematical model is a function connecting two variables representing real world quantities: the function shows how one variable changes with the other. A model rarely matches real-life exactly – there are often too many factors to take into account, some of them possibly unknown. To take a simple example: can a mathematical model predict the exact spot a ball will land when thrown through the air? A simple model, which treats the ball as a projectile with gravity being the only force acting on it, may give a good approximation. A better model would take air resistance into account, but this also needs to involve the characteristics of the ball. Then the model could be further refined by taking into account the velocity of the wind, and the profile of the ground.

You should understand the basic modelling cycle: propose a model; test the model and, if necessary, refine it to improve it; then test it again. Continue until the model reflects the real world situation to a desired accuracy.

How might exam questions approach mathematical modelling?

- A function is suggested to represent a real world situation. This might include constants which you have to evaluate by substituting given values into the function. You may be asked to suggest a reasonable domain for the function.

- The question could ask you to sketch the graph of the function, and then ask you to draw conclusions from its shape or significant points or features.

- You might have to substitute real world values into the function, and comment on how good the model is; you could be asked to suggest reasons why the model isn't accurate.

- A second function could be proposed as a better model – again, you might have to test values and comment on how much more accurate the model is.

- You could be asked to use the model to make predictions. Be aware that *extrapolation* – where predictions are made beyond the range within which the model has proved accurate – can give false predictions.

A researcher is using records of the estimated number of trees on an island to devise a model for its deforestation. The number of trees at different dates is shown in the following table:

Year	1820	1880	1930	1950	1970	1980	1990	2010
Trees (thousands)	3000	2900	3200	2450	1800	1340	1000	740

    (a)   Draw a graph of the data with the year on the x-axis and the number of trees on the y-axis. Use scales of 20 years/cm on the x-axis and 250,000 trees/cm on the y-axis.

    (b)   Why does the graph suggest deforestation may have begun in 1930?

The researcher devises a mathematical model for the deforestation of the form $N = ab^t$, where a is the constant 3.2×10^6, t is the number of years since 1930, N is the number of trees, and b is a constant to be found.

    (c)   Given that the number of trees in 2010 is known to be accurate, calculate the value of b.

    (d)   Assuming the model is valid, in what year would the number of trees be expected to fall below 500,000.

    (e)   Why might the model no longer be valid in 2050?

(a)

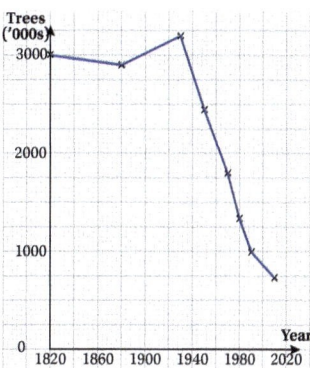

(b)   Number of trees was fairly constant until 1930, then fell continuously.

(c) $7.4 \times 10^5 = 3.2 \times 10^6 \times b^{80}$

    $b^{80} = \dfrac{7.4 \times 10^5}{3.2 \times 10^6} = 0.23125$

    $b = \sqrt[80]{0.23125} = 0.982$

(d) $5 \times 10^5 = 3.2 \times 10^6 \times 0.982^t$

    $t = 102.2$ years

    $\therefore$ would fall below 500,000 during 2032

(e)   Climate change may affect deforestation, or government policies to prevent complete clearance.

It is important to note that in both the table and on the graph, the number of trees is measured in thousands.

(c) The numbers look a bit scary, but keep going and the answer seems reasonable.

In part (d), remember that t is the number of years since 1930. The exact value of t is 101.4 years, but 102.2 is allowable.

Plenty of possible answers in part (e).

We shall now look in more detail at the various types of functions which could be used for modelling.

2.4 Linear Functions

In a linear function, the function increases (or decreases) at a constant rate. Its graph is a straight line.

What you need to know about straight line graphs

Formulae

The gradient between two points (x_1, y_1) and (x_2, y_2) is $\dfrac{y_2 - y_1}{x_2 - x_1}$

Parallel lines: $m_1 = m_2$

Perpendicular lines: $m_1 = -\dfrac{1}{m_2}$

Midpoint of two points is $\left(\dfrac{x_1 + x_2}{2}, \dfrac{y_1 + y_2}{2}\right)$

Distance between two points is $\sqrt{(x_2 - x_1)^2 + (y_2 - y_1)^2}$

Gradient: The *gradient* of the line is its "steepness." A gradient of 3 means that y is increasing 3 times faster than x. The gradient is calculated by choosing two points and dividing the change in y by the change in x, often remembered as $\dfrac{\text{rise}}{\text{run}}$.

Horizontal lines have gradient 0. Vertical lines have an infinite gradient. Lines angled from bottom left to top right have positive gradients, others have negative gradients.

Midpoint, distance between two points: The midpoint of two points can be found by calculating the x-coordinate halfway between the x-coordinates of the two points, and the same for the y-coordinate. The distance between two points is calculated using Pythagoras' Theorem.

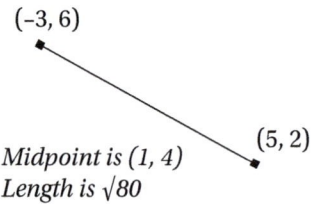

Midpoint is (1, 4)
Length is $\sqrt{80}$

Although the formulae are shown in the box, this means there are a lot of formulae to remember. It is often better to draw a sketch and work from that.

Drawing a line from its equation:

A useful line to remember is the one which passes through $(a, 0)$ and $(0, a)$. Its equation is $x + y = a$.

- If the equation is of the form $y = mx + c$, substitute 2 or 3 values for x and work out the corresponding y values.

- If the equation is of the form $ax + by = c$, it is easier to put x equal to 0 and work out y, then put y equal to 0 and work out x. This gives the two points where the line crosses the axes.

- To *sketch* the graph of $y = mx + c$, remember that c is the y-intercept and m is the gradient.

Working out the equation of a line from the graph:

1. Calculate the gradient.

2a. If using the first formula, replace m with the gradient, then substitute a point for x and y.

3a. Calculate c and then put this back into the equation.

2b. If using the second formula, replace m with the gradient then substitute the point for x_1 and y_1.

3b. Rearrange and simplify to get the equation.

There are two formulae you can use:

$y = mx + c$

$y - y_1 = m(x - x_1)$

The points P, Q have coordinates P(3, 0), Q(-3, 7). Find the equation of the line which is perpendicular to PQ and passes through P. Give your answer in the form $ax + by + c = 0$, where a, b and c are integers.

Gradient of PQ $= \dfrac{7-0}{-3-3} = -\dfrac{7}{6}$

$\therefore$ Gradient of perpendicular line to PQ $= \dfrac{6}{7}$

Equation of line is $y - y_1 = m(x - x_1)$

$$y - 0 = \dfrac{6}{7}(x - 3)$$

$$7y = 6x - 18$$

So the line has equation $6x - 7y - 18 = 0$

Whenever you need to find the equation of a straight line your first thought should be: "I need a point, I need a gradient."

I prefer to use this formula when finding the equation of the line because all the numbers are substituted in one go.

It's unlikely you will get an exam question purely on the equations of straight lines. But you will need to use your knowledge of straight line techniques in many situations; for example, mathematical modelling, equations of tangents to curves, Voronoi diagrams.

Linear models:

Examples:

Real world situation	Possible linear model
Distance d km travelled in t hours in a car going at a constant speed.	$d = 80t$
Cost £ C of printing n leaflets.	$C = 5 + 0.02n$
Height h metres reached by a balloon t seconds after being launched from the top of a building.	$h = 20 + 3.5t$

In the first example, a graph of d against t would go through the origin. This is because at time 0 hours, the distance travelled is 0 km. However, in the other examples something different is happening. The y-intercept on the second graph is £5; this represents the initial cost, in the case the cost of setting up the print run. After this, each leaflet will cost £0.02 to print. Similarly, the 20 m in the third example must be the height of the building – in other words, the height of the balloon above the ground at $t = 0$.

Piecewise models:
Linear models can also be *piecewise*, and in such cases are formed by several sections of straight line rather than one continuous line with a fixed gradient.

For example, look at this graph which shows the cost, in crowns, of posting parcels of different weights in the country of Alluria:

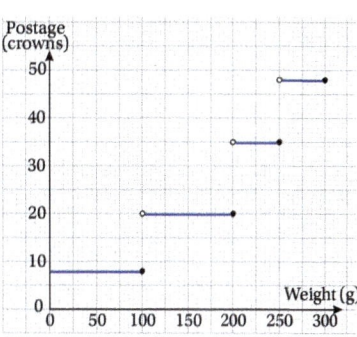

A parcel weighing 60 g would cost 8 crowns to post, as would a parcel weighing 80 g. But a parcel weighing 270 g would cost 48 crowns. What about one which weighs exactly 200 g? The black dots on the lines are inclusive, the white dots exclusive – so it would cost 20 crowns. For a graph like this, it wouldn't be necessary to find the equations of each of the line segments.

Piecewise models can also be formed of connected lines.

Here's a typical example:

Example: Two venues offer different pricing models. For an event at venue A there is an initial $150 fee, plus $10 per person. At venue B an event costs $100 for the first 15 people, and then an additional $20 per person.

(a) Complete the following table for each venue:

	10 people	15 people	20 people	30 people	40 people	50 people
Venue A	$250				$550	
Venue B			$200			

(b) On a single diagram, plot the points to create graphs for the cost of both venues.

(c) Suggest an equation connecting cost $ C with number of people n for venue A.

(d) Show that the equation for the second section of the graph for venue B is $C = 20n - 200$.

(e) For how many people do both venues cost the same?

Solution:

(a)

	10 people	15 people	20 people	30 people	40 people	50 people
Venue A	$250	$300	$350	$450	$550	$650
Venue B	$100	$100	$200	$400	$600	$800

(b)

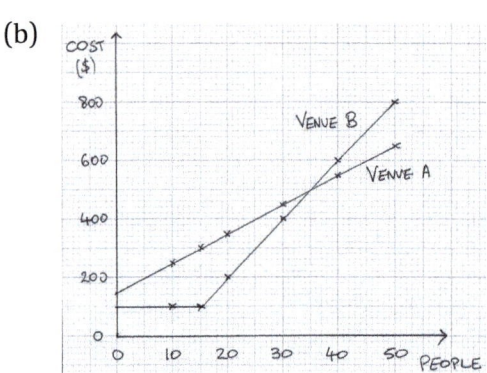

(c) y-intercept = 150, gradient = 10 $\therefore$ Equation is $C = 150 + 10n$

(d) Test two points: For (20, 200), $200 = 20 \times 20 - 200$

For (40, 600), $600 = 20 \times 40 - 200$

The two points satisfy the equation, so the equation is valid.

(e) 35 people ($500 for both models)

The graph shows that the pricing model for venue A is a simple linear one, but for venue B is piecewise linear.

Linear Functions: Practice Exercise

1. Each of the points $(1, -1)$, $(-1, 1)$, $(1, 0)$ and $(-2, 1)$ lies on one of the lines with the following equations:

    Line A: $y - 2x = 3$

    Line B: $y = 4x - 4$

    Line C: $3y = x - 4$

    Line D: $2y + x = 0$

    Which point lies on which line?

2. (a) Find the equation of the line passing through $(2, 4)$ and $(4, 10)$.

    (b) Find the equation of the line passing through $(1, 5)$ and $(4, -1)$.

    (c) Find the point of intersection of the two lines.

3. A line L intersects the x-axis at A$(6, 0)$ and the y-axis at B$(0, 2)$.

    (a) Write down the coordinates of the midpoint of AB.

    (b) Calculate the gradient of L.

    (c) Find the equation of the line parallel to L passing through the point $(6, 2)$ in the form $ax + by = c$.

    (d) Calculate the length of AB.

    (e) Calculate the area of triangle OMA. *A diagram will help.*

4. The height h metres of a balloon launched from the top of a building is modelled by the function $h = 12 + 4.5t$, where t is the time in seconds after launch.

    (a) How tall is the building?

    (b) What is the height of the balloon after 8 seconds.

    (c) The balloon bursts when it reaches a height of $105\,\text{m}$. How many seconds after launch did it burst, giving your answer to the nearest second?

2.5 Quadratic Functions

Quadratic functions are of the form $ax^2 + bx + c$ where $a \neq 0$.

What you need to know about quadratic graphs:

Shape of a quadratic graph: All quadratic graphs are parabolas, the sign of a determining "which way up." If the equation of the graph is written in the form $y = ax^2 + bx + c$ we can say which way up the graph is and where the y-intercept is. For example, the graph of $y = x^2 + 3x - 4$ cuts the y-axis at $(0, -4)$ and is in the shape of a U.

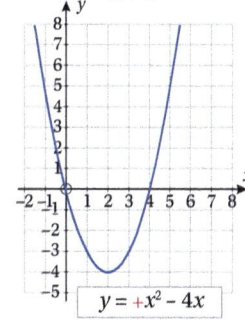

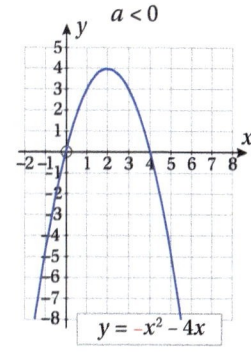

Line of symmetry: A quadratic graph is always symmetrical about the vertical line passing through the vertex (turning point), a fact which can often be used when answering

47

questions about the graph. The equation of this line is $x = -\dfrac{b}{2a}$. This is why, for both graphs illustrated, the line of symmetry is $x = 2$.

Vertex: The vertex can be found using the maximum or minimum function on your GDC. It is also easy to find using the equation. For example, take the graph of $y = 2x^2 - 4x + 1$. In this case, $a = 2$ and $b = -4$, so the line of symmetry is $x = -\dfrac{(-4)}{2 \times 2} = 1$. This must be the x coordinate of the vertex and, substituting in to the equation, this gives $y = -1$. So the vertex is at $(1, -1)$. We can also tell that this is a minimum because $a > 0$.

Intercepts, zeroes and roots: The y-intercept of a quadratic graph can be found by putting $x = 0$ into the equation. The x-intercepts can be found using the polynomial equation solver, or by using the "zero" function on your GDC (although this method will only give you one intercept at a time).

> You do also need to be able to factorise quadratic expressions. This is part of your "prior learning." If you need a refresher, I have included the essential knowledge on the website.

Example: Find all the axis intercepts on the graph of $y = 2x^2 - 5x + 3$.

Solution: y-intercept when $x = 0$, so $y = 0 - 0 + 3 = 3$.

x-intercepts when $2x^2 - 5x + 3 = 0$.

Solutions: $x = 1$ and 1.5 (GDC), or by factorisation $y = (2x - 3)(x - 1)$

$\therefore$ Axis intercepts are $(0, 3)$, $(1, 0)$, $(1.5, 0)$.

Note the terminology: the x-intercepts are the *zeroes* of the function and also the *roots* of the equation.

Let $f(x) = 4x^2 + bx + c$.

    (a)   Given that the y-intercept is $(0, -20)$, write down the value of c.

    (b)   Given that the line of symmetry is $x = 0.25$, find the value of b.

    (c)   The x-intercepts are at $(h, 0)$ and $(k, 0)$ where $k > h$. Find the value of $k - h$.

(a) $c = -20$	*Yes, (a) is as simple as it seems!*
(b) $-\dfrac{b}{2a} = 0.25$ so $-\dfrac{b}{8} = 0.25 \Rightarrow b = -2$	*In part (b) we just use the equation of the line of symmetry "in reverse" to find the value of b.*
(c) Equation is $y = 4x^2 - 2x - 20$.	*Part (c) breaks down to "find the solutions of the quadratic equation and subtract them." Make sure you understand the interconnectivity of intercepts, roots and zeroes.*
Roots are -2 and 2.5 (GDC)	
So $h = -2$ and $k = 2.5$, and $k - h = 4.5$.	

Note that if you know the line of symmetry (or the coordinates of the vertex) of a quadratic and one of the roots, then you can use symmetry to find the other root. For example, if the vertex is at $(3, 6)$ and one of the roots is $x = -1$ then, by symmetry, the other root must be $x = 7$.

Here's another example where a question uses the symmetry of a quadratic graph:

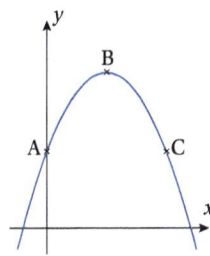

The diagram shows the graph of the quadratic function $f(x) = ax^2 + bx + 4$ with vertex B $(3, 10)$.

(a) Write down the coordinates of point A.

(b) Use symmetry considerations to write down the coordinates of a third point C.

(c) Use the coordinates of B and C to write down two equations in a and b, and hence calculate their values.

Quadratic models: The techniques used in the previous section can be used to find solutions when real world problems are modelled as quadratic functions. Let's consider the following model, for a ball hit through the air, and some of the questions which might be posed.

The height of the ball above the ground is modelled by the function $h(t) = 2.1 + kt - 5t^2$, $0 \le t \le c$ where h is in metres and t is the time in seconds after the ball has been hit.

At what height was the ball hit?

When $t = 0$, $h = 2.1$, so the ball was hit at a height of 2.1 m.

If the ball is at a height of 12.1 m after 1 s, find k.

Substituting, $12.1 = 2.1 + k - 5 \Rightarrow k = 15$.

When does the ball hit the ground?

We want $h = 0$, so solve $0 = 2.1 + 15t - 5t^2$. GDC gives $t = -0.13$ or 3.13. Since $t > 0$, the ball hits the ground when $t = 3.13$ s.

State, with a reason, the value of c.

The model is no longer valid once the ball hits the ground, so $c = 3.13$.

We could also find t using $\frac{-b}{2a}$ since the highest point is on the line of symmetry. Confirm that this gives $t = 1.5$ s.

Find the maximum height of the ball given that it reaches this height when $f'(t) = 0$.

$f'(t) = 15 - 10t$, so $f'(t) = 0$ when $t = 1.5$. At $t = 1.5$, $h = 13.35$ m.

Quadratics: Practice Exercise

1. Solve these quadratics using factorisation, rearranging into the form $ax^2 + bx + c = 0$ as necessary. Also solve directly on the GDC using either intersections of graphs, or roots of equations.

   (a) $x^2 - x + 2 = x + 17$

   (b) $x = 5 - \frac{4}{x}$

   (c) $2x^2 - 5x - 12 = 0$

   (d) $20x^2 + 160x + 140 = 0$

2. Solve, giving answers the form $a + b\sqrt{c}$, where $a, b, c \in \mathbb{Q}$

   (a) $2x^2 + 4x - 1 = 0$

(b) $(x - 1)^2 = 6 - (x + 2)^2$

3. The area of the shape on the right is 33 cm².

    Set up an equation in x and solve to find the dimensions of the rectangle.

    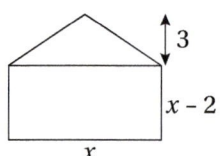

4. The function $f(x) = 4x^2 - kx + 25$ where $k > 0$ has its vertex on the x-axis. Find the value of k, the coordinates of the vertex, and the point of intersection of the graph of f with the line $y = 1$.

5. The graph shows the function $f(x) = 8x - ax^2$. The x-intercepts are at $(0, 0)$ and $(4, 0)$. Find the value of a, the equation of the line of symmetry, and the coordinates of the vertex.

    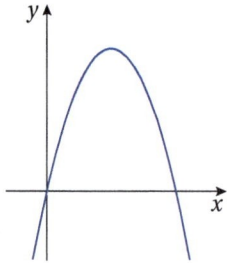

2.6 Exponential Functions

See page 11 for an introduction to exponents.

An exponent is a power, so exponential functions are based on $f(x) = a^x$. Note that the exponent is x, unlike a function such as $f(x) = x^2$ where the exponent is a constant. An important feature of *every* exponential graph is the presence of a horizontal asymptote, which means that an exponential function tends towards a particular value – but never reaches it.

What you need to know about exponential graphs:

Let's look at the graphs of these three functions: $f(x) = 2^x$, $f(x) = 3 \times 2^x$, $f(x) = 3 \times 2^x - 1$.

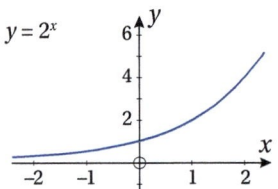 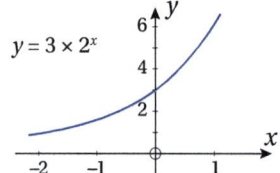 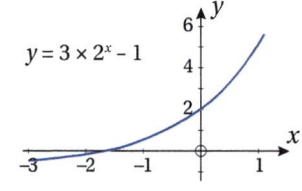

Remember that negative powers do *not* result in negative numbers. That is why the graph of $y = 2^x$ never crosses the x-axis. Also note that *all* graphs of equations of the form $y = a^x$ have the same shape, and all have a y-intercept $(0, 1)$.

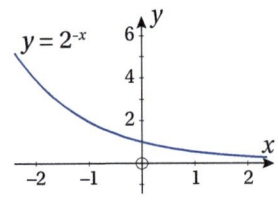

$y = 2^x$ passes through $(0, 1)$ because $2^0 = 1$. For all values of x the graph climbs higher at an ever-increasing rate, displaying "exponential growth." For $x < 0$, remember that $2^{-1} = \frac{1}{2}$, $2^{-2} = \frac{1}{4}$ and so on. Each time x decreases by 1, the distance to the x-axis halves – so the graph creeps down towards the axis, but never reaches it. Thus the x-axis, whose equation is $y = 0$, is an *asymptote*.

The graph of $y = 3 \times 2^x$ looks pretty much the same. The asymptote is $y = 0$ and it still climbs exponentially towards infinity. The key difference is that the y-intercept is $(0, 3)$ since $3 \times 2^0 = 3 \times 1 = 3$. To draw the third graph we simply subtract 1 from all the y values in the second graph. Thus the intercept is $(0, 2)$ and, critically, the asymptote is now $y = -1$. We could test this by putting in a large negative number for x. For example, $3 \times 2^{-10} - 1 = -0.99707$, which is very close to -1.

If we replace x with $-x$ the effect is to reflect the graphs in the y-axis. For example, the graph of $y = 2^{-x}$ is shown in the diagram. These graphs represent exponential decay – we shall see more in the next section on modelling with exponential graphs.

Questions may use different values of a ($a > 0$) in functions of the form $f(x) = k\,a^x + c$ and $f(x) = k\,a^{-x} + c$, but you can also expect specifically to see functions of the form $f(x) = k\,e^{rx} + c$. 'e' is an irrational number, a constant with value $e \approx 2.718$. Although it crops up in almost every area of mathematics, our particular interest here is with e as the base of an exponential function. $y = e^x$ has the valuable property that the gradient of the graph at any point is equal to the function value (ie the y value) at that point; it is therefore often found in modelling functions related to population growth, where the rate of change of the population is directly related to the number of people in the population.

Exponential models: There are many different situations which exhibit exponential growth or decay, best illustrated with a few problems

> A group of ten monkeys is introduced to a zoo. After t years the number of monkeys, N, is modelled by $N = 10e^{0.3t}$.
>
> (a) How many monkeys are there after 3 years?
>
> (b) How many complete months will it take for the number of monkeys to reach 50?

(a) $N = 10\,e^{0.3 \times 3} = 24.596...$

    $\therefore$ There are 24 monkeys

(b) $50 = 10\,e^{0.3t}$

    $5 = e^{0.3t}$

    $\ln 5 = \ln e^{0.3t}$

    $\ln 5 = 0.3t$

    $t = 5.36...$ years $= 64.38$ months

    $\therefore$ There will be 50 monkeys after 65 months

An important point to note in the answers to this question is that they are not rounded in the normal way. In part (a) there are only 24.596 monkeys after three years, so the 25th monkey hasn't appeared yet! The answer would still be 24 even if the calculation came to 24.99.

And in part (b), although 64.38 rounds to give 64, there won't be 50 monkeys after 64 months. In terms of complete months, the answer is 65.

The value of t could also be found by solving $5 - e^{0.3t} = 0$ on your GDC.

A typical example where the asymptote comes into play is illustrated by a function modelling the cooling of a liquid. Suppose a flask of boiling water is placed in a room where the temperature is 18°C and left to cool. It will start cooling quite quickly then, as the water temperature approaches room temperature, the rate of cooling slows. An exponential function can be used to model this situation, but because of the asymptote it appears that the water will never quite reach 18°C – in practice it will get there in the end.

Example: The graph shows the temperature T °C of a cup of boiling water at a time m minutes since the water was poured. Room temperature is 18°C.

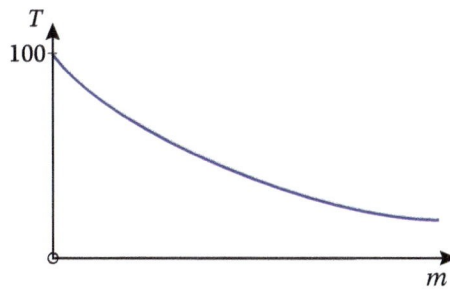

The situation is modelled by the function $T = p + qa^{-m}$.

(a) Write down the value of p.

(b) Find the value of q.

In part (b), remember that $a^0 = 1$.

(c) After 2 minutes the water temperature is 75 °C. Use this fact to show that $a = 1.2$ to 1DP.

(d) Find the time taken for the water to reach 30 °C. Give your answer in minutes and seconds, correct to the nearest second.

Solution:

(a) The water temperature approaches room temperature, so $p = 18$.

(b) When $m = 0$, $T = 100$. $100 = 18 + qa^0 \Rightarrow q = 82$.

(c) Substituting, $75 = 18 + 82 \times a^{-2} \Rightarrow a = 1.1994... = 1.2$ to 2SF

(d) Solve $30 = 18 + 82 \times 1.2^{-m} \Rightarrow m = 10.54$ minutes

0.54 minutes $= 32.4$ s, so $m = 10$ m 32 s

In some exponential models, the function increases to or decreases from the asymptote. For example, the model for a mobile phone battery whilst being charged might be $C(t) = 4.5 - 3 \times (2^{-t})$, $t > 0$, where C is the charge in volts/cell, and t the time in minutes the battery has been charging. Here's the graph:

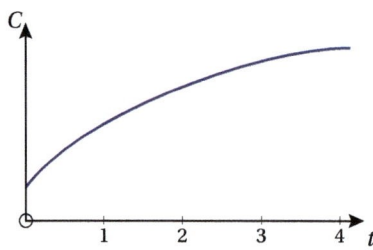

The y-intercept is $4.5 - 3 \times 2^0 = 1.5$, the charge level at the start. And to find the asymptote, just substitute a large value for t, say 100. This gives $C = 4.5 - 3 \times 2^{-100} \approx 4.5$, the level of maximum charge.

Half-life: One of the features of functions of the form $f(x) = a^x$ is that the value of the function multiplies by the same amount in equal intervals. Consider the function $f(x) = 3.5^x$:

x	1	3	5
$f(x)$	3.5	42.875	525.21875

Now $42.875 \div 3.5 = 12.25$ and $525.21875 \div 3.5 = 12.25$. In other words, every time x increases by 2, the function multiplies by 12.25.

Let's look at it another way: when $x = 1$, $f(1) = 3.5$. When does the function have value 7? We need to solve $3.5^x = 7$, giving $x = 1.553....$ So the function doubles in value when x increases by $0.553...$, and this will be true whichever value of x you choose as a start point.

When the function represents decay, rather than growth (ie when the exponent is negative) this leads to the concept of a *half-life*. When a mass, or level of radioactivity, is decaying exponentially, the length of time it takes to reduce by half is the half-life. The general formula for half-life calculations is $m_t = m_0 e^{-kt}$ where m_t is the mass or activity level at time t, m_0 is the initial mass or activity, and k is a constant dependent on the half-life. If we substitute $m_t = 0.5$ and $m_0 = 1$, we find that $k = -\dfrac{\ln 0.5}{\text{half-life}}$.

Try these quick practice half-life questions:

1. After 56 days the activity of a sample of phosphorus-32 has decreased from 400 Bq to 25 Bq. What is the half-life, and what is the formula for the activity remaining after t days?

2. The half-life of thorium-227 is 19 days. How many days are required for 80% of a sample to decay?

3. A rock once contained 3.5 mg of uranium-238, but now only contains 0.25 mg. Given that the half-life of uranium-238 is 4.5×10^9 years, how old is the rock?

Answers

1. 14 days, $a = 400\,e^{-0.0495t}$

2. 44.1 days

3. 1.7×10^{10} years

Full working on the website

2.7 Reciprocal functions

Functions such as $f(x) = \dfrac{a}{x}$ and $f(x) = \dfrac{a}{x^2}$ look simple enough, but since division by zero will always result in an asymptote on a graph, all functions which contain a power of x as the denominator will also have asymptotes.

Let's consider the function $f(x) = \dfrac{12}{x}$ and draw up a table of values:

x	-6	-4	-2	-1	0	1	2	4	6
$\dfrac{12}{x}$	-2	-3	-6	-12	—	12	6	3	2

And here's the graph:

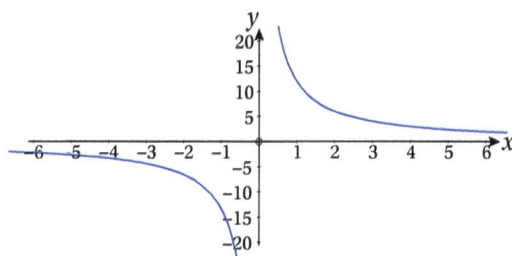

The graph of $y = \dfrac{6}{x^2}$ is similar except that all the y values are positive.

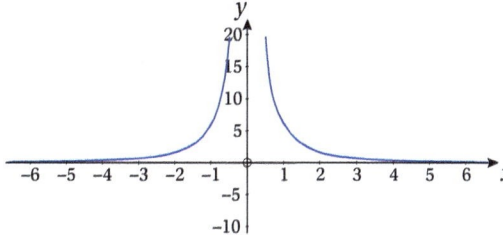

But this is not necessarily true when the denominator contains something more complicated than a power of x. For example, $y = \dfrac{1}{x^2 + 1}$ has no asymptote since $x^2 + 1$ cannot equal 0.

Generally, graphs of functions $f(x) = ax^{-n}$ all have the y-axis as a vertical asymptote; for odd powers of n they look like the first graph, and for even powers of n they look like the second graph.

Direct and inverse variation: Quantities connected by the equation $y = ax^n$ where $n \in \mathbb{Z}$ have a proportional relationship. The types of proportion you are likely to meet are named in the following table.

$y = ax$	Direct proportion
$y = ax^2$	Square proportion
$y = ax^3$	Cubic proportion
$y = \dfrac{a}{x}$	Inverse proportion
$y = \dfrac{a}{x^2}$	Inverse square proportion
$y = \dfrac{a}{x^3}$	Inverse cubic proportion

Questions involving proportional quantities are solved in the same way:

1. Write down the relevant equation as in the table.

2. Substitute given values to find the constant of proportionality.

3. Rewrite the equation, substituting the constant.

4. Answer any further questions.

The weight of an object is inversely proportional to the square of its distance from the centre of the Earth. Assume the Earth is a sphere with radius 6300 km. A small communications satellite weighs 1800 kg on the Earth's surface. How much will the satellite weigh when in orbit 2500 km above the Earth's surface?

$$w = \frac{k}{d^2}$$

When $d = 6300$, $w = 1800$

$\therefore\ 1800 = \dfrac{k}{6300^2} \Rightarrow k = 7.1442 \times 10^{10}$

Thus $w = \dfrac{7.1442 \times 10^{10}}{d^2}$

In orbit, $d = 6300 + 2500 = 8800$

So $w = \dfrac{7.1442 \times 10^{10}}{8800^2} = 923$ kg.

The key to a question like this is to start by writing down the proportionality equation – everything else then follows.

Once in orbit, the distance of the satellite from the centre of the Earth will be the Earth's radius plus the orbital height.

2.8 Cubic Functions

A cubic function is of the form $ax^3 + bx^2 + cx + d$.

What you need to know about cubic graphs: The graph of the cubic function $f(x) = ax^3 + bx^2 + cx + d$ starts down the "bottom left" of the 3rd quadrant and ends up at the "top right" of the 1st quadrant if $a > 0$. If $a < 0$ it will start up at the top left and end at the bottom right. This means that there must be at least one zero since the graph will definitely cross the x axis. A cubic graph can have a couple of turning points, so there could be up to three zeroes.

The diagram shows a typical cubic graph where $a > 0$ and there are three zeroes.

Excuse the informal terminology – but you know what I mean!

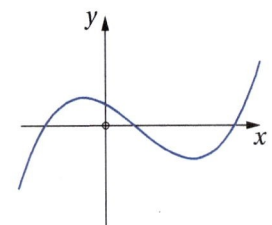

Example: The graph of $f(x) = ax^3 + bx^2 + cx + 6$ passes through the points $(1, -6)$, $(-1, 12)$ and $(2, -12)$. Write down three equations in a, b and c and hence find the roots of $f(x) = 0$.

Solution: Substituting for x and y gives:

$$-6 = a + b + c + 6 \implies a + b + c = -12$$

$$12 = -a + b - c + 6 \implies -a + b - c = 6$$

$$-12 = 8a + 4b + 2c + 6 \implies 8a + 4b + 2c = -18$$

Solve simultaneously (GDC): $a = 2$, $b = -3$, $c = -11$

Thus $f(x) = 2x^3 - 3x^2 - 11x + 6$

$f(x) = 0 \implies x = 0.5, -2, 3$ (GDC)

2.9 Trigonometric Functions

The graphs of $\sin x$ and $\cos x$: The diagram shows a circle with radius 1 (a *unit circle*). A line is drawn from the centre to a point on the circumference, and this forms angle θ with the x-axis. The x-coordinate of the point is defined as the *cosine* of the angle ($\cos\theta$) and the y-coordinate as the *sine* ($\sin\theta$).

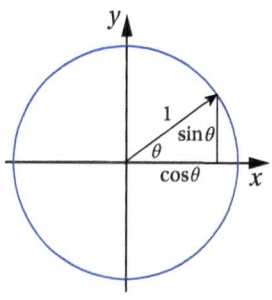

These definitions will apply as the line rotates full circle, giving the sin and cos for all angles from 0° to 360°. When these are plotted as graphs, we get the following:

These graphs can, of course, be extended to show the sin and cos for *all* angles.

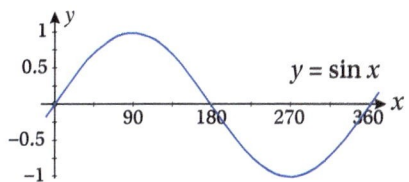

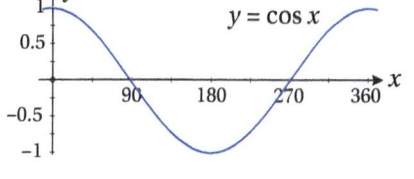

Points to note:

- The range of both functions is $-1 \leq f(x) \leq 1$, and both have an *amplitude* (height of the wave from the centre line) of 1.
- $\sin x > 0$ for angles between 0° and 180°.
- $\cos x > 0$ for angles between 0° and 90°, also between 270° and 360°.
- Both functions have a *period* (ie repeat themselves) every 360°.
- The graph of $\cos x$ is exactly the same as the graph of $\sin x$ shifted left by 90°.

Radians: Generally, angles are measured in degrees. But when dealing with trigonometric functions (often called *sinusoidal functions*) a different angular measure, *radians*, can be used – and **must** be used when calculus is involved. The number of radians in a complete circle is 2π, and so the conversion is 360° = 2π radians or, easier to use, 180° = π radians.

If a question is in degrees, the degrees symbol will always be used.

Solving trigonometric equations: Use the arcsin function on your GDC to solve the equation $\sin x = 0.5$, and the solution is $x = 30°$. However, you can see from the graph above that there is a second angle with the same value; similarly, there is a second angle on the unit circle with the same height. Symmetry considerations lead us to the second

For $\cos x = 0$ and $\cos x = 1$ there are two solutions, and for $\cos x = -1$ there is one solution.

See website for details of this method.

Answers

(a) 40.7°, 139.3°

(b) 60°, 300°

(c) 240.5°, 299.5°

(d) 150°, 210°

solution which is $x = 150°$. In other words, there are always two solutions (in the range $0° \le x \le 360°$) to the equation $\sin x = k$ where $-1 \le k \le 1$. The only exceptions are $\sin x = \pm 1$, each with just one solution, and $\sin x = 0$, which has the grand total of 3 solutions. Look at the graph to see why. All of this is true for $\cos x = k$, although the exceptions are different.

To solve trigonometric equations such as these, you can either use the symmetries of the graphs or the unit circle, or you can draw the graph on your GDC and use the normal methods of solution.

Try these – they all have two solutions. Answers to 1DP.

(a) $\sin x = 0.652$ (b) $\cos x = 0.5$ (c) $\sin x = -0.87$ (d) $\cos x = -\dfrac{\sqrt{3}}{2}$

🖩   Answers may be required in radians – always ***start*** by checking which mode your calculator is in.

Trigonometric identities: Trigonometric expressions can often be simplified by using a pair of useful identities:

- For any angle θ, $\sin^2\theta + \cos^2\theta = 1$ and $\dfrac{\sin\theta}{\cos\theta} = \tan\theta$.

Show that the solutions between 0° and 360° are 193° and 347°.

For example, we could show that the equation $2\cos^2\theta + 4\cos\theta\,\tan\theta = 1$ can be rewritten as $2\sin^2\theta - 4\sin\theta - 1 = 0$. This is done by replacing $\cos^2\theta$ with $1 - \sin^2\theta$ and $\tan\theta$ with $\dfrac{\sin\theta}{\cos\theta}$. The resulting equation can then be solved either graphically or as a quadratic.

It is also worth remembering that $\sin(-\theta) = -\sin\theta$ whereas $\cos(-\theta) = \cos\theta$.

Trigonometric models: In many situations where some sort of circular motion is involved, real world situations may best be modelled by a sin or a cos function. For example, the height of a person on a big wheel at different times; the height of the tide at different times (which depends on the circular motion of the moon); voltage levels generated by an alternating current generator (where a magnetic rotor is turning at high speed). Typically, to model such situations, the functions will be of the form $f(x) = a\sin(b(x - c)) + d$ or $f(x) = a\cos(b(x - c)) + d$. The values of a, b, c and d have specific counterparts in the real world situation.

The only difference between using sin and cos is whether the start point (at $t = 0$) is in the middle of the wave, or at the high or low point.

- a is the *amplitude* which is the distance from the centre line to the top or the bottom of the wave.

- b is connected to the *period*. $\sin x$ and $\cos x$ both have a period of 360°. If b is, say, 2, the wave will have double the frequency; that is, it will complete one cycle every 180°. Thus the period is $\dfrac{360°}{b}$.

See page 19 for more details.

- c shifts the wave horizontally – such a translation is known as a *phase shift*.
- d shifts the centre line (the *principal axis*) of the wave up or down.

For example, the graph of $f(x) = 3\sin(6(x - 20))° + 2$ will have a period of 60°, a principal axis of $y = 2$, an amplitude of 3, and a phase shift of 20°, from which it is just one step to calculate that its minimum and maximum values are 2 ± 3, giving a range of $-1 \le f(x) \le 5$. Try sketching the graph, and then checking it on your GDC.

The following example shows how the function can be deduced from the equation of its graph.

Example: The graph of $f(x) = p\cos(qx)° + r$, for $-30° \le x \le 30°$, is shown below. A is a maximum point, B is a minimum point.

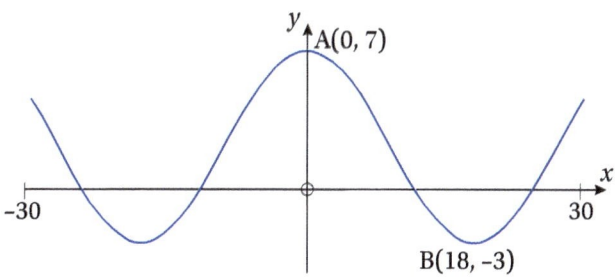

(a) Find the value of

(i) p

(ii) q

(iii) r

(b) The equation $f(x) = k$ has exactly two solutions. Write down the value of k, given that $k < 0$.

(c) For the values found in part (a), solve $p\cos(qx) + r = 6$, $0 \le x \le 18$.

Solution: *The key to answering this question lies in the two turning points. Firstly, the difference between the y-coordinates is 10, so the amplitude of the wave is 5, and this leads us to the value of p.*

Looking at the two x-coordinates, we can deduce that the period (the length of one wave) is 2 × 18 = 36. This gives us the value of q (see the note on the period, above).

Finally, looking at the y-coordinates again, we can see the principal axis of the wave is y = 2, and this gives us the value of r.

(a) (i) $p = 5$ (ii) $q = \dfrac{360}{36} = 10$ (iii) $r = 2$

To answer questions like that in part (b), all you have to do is work out where a horizontal line added to the diagram would cut the graph in only two places. The only possibilities, given that k < 0, is a line passing through the minimum points.

(b) $k = -3$

(c) $x = 3.687°$ (GDC)

Here's another graph with equation of the form:

$$y = A\sin\left(\tfrac{\pi}{B}(x - C)\right) + D$$

Watch the video online to discover how to find the values of A, B, C and D.

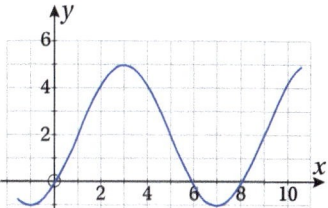

Any real world environment involving waves can be modelled by a sinusoidal function, examples being: height of tide over time (because of the circular motion of the moon), length of the day over a year, sound intensity.

The depth $h(t)$ metres of water at the entrance to a harbour t hours after midnight on a particular day is given by $h(t) = 5 + 3.5\sin\left(\frac{\pi}{6}(t-1)\right)$, $0 \le t \le 12$.

(a) Explain the relevance of the domain.

(b) Sketch the graph of h, marking any axis intercepts and the principal axis.

(c) Write down the minimum and maximum depths of the water.

(d) A boat is moored in the harbour overnight. If it requires a water depth of 2.1 m to stay afloat, calculate the time by which it should set sail, giving your answer to the nearest minute.

(a) The function gives the height of the tide from midnight to midday.

(b)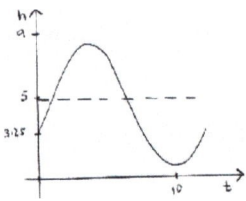

The function uses radians since there are no degrees signs – and also the presence of π is a pretty strong clue! So make sure your GDC is set to radians before you do anything else.

(c) 1.5m, 8.5m

(d) $5 + 3.5\sin\left(\frac{\pi}{6}(t-1)\right) = 2.1 \Rightarrow t = 8.665$ (GDC)

$0.665 \times 60 = 39.9$

∴ Set sail by 0840.

2.10 Logistic Functions

Logistic functions are of the form $f(x) = \dfrac{L}{1 + Ce^{-kx}}$, $L, k, C > 0$. They provide a model where the growth of a quantity is prevented from passing a limiting value.

What you need to know about the graph of a logistic function:

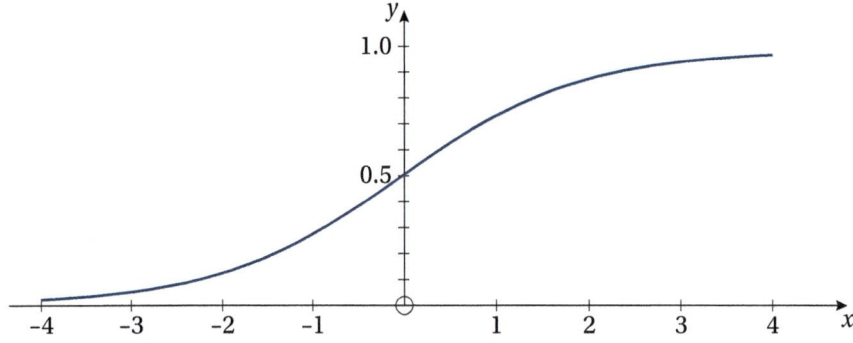

The diagram shows the graph of the "standard" logistic function, $f(x) = \dfrac{1}{1 + e^{-x}}$. All logistic functions show the same general shape – rate of growth increasing to an inflexion point, then decreasing until the value approaches the limiting value, represented by the upper horizontal asymptote.

So, what effect do the values of L, C and k have on the graph?

- L is the limiting value. Thus the graph will always have asymptotes $y = L$ and $y = 0$. Note that the y-coordinate of the point of inflexion is midway between the two asymptotes.

- Increasing the value of k "squeezes" the middle part of the growth into a shorter period; in other words, the gradients are greater in the middle section of the graph. But the point of inflexion remains unchanged.

- C changes the y-intercept. When $x = 0$, $y = \dfrac{L}{1 + C}$. Larger values of C have the effect of shifting the main part of the growth towards the second half of the graph, or shifting the graph horizontally. Also, the y-coordinate of the point of inflexion has value $C \div 2$.

The effect of changing these values can be seen in the three graphs below.

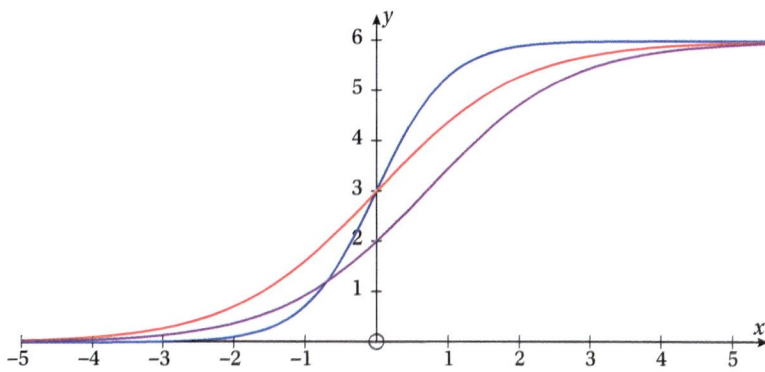

The red graph has equation $y = \dfrac{6}{1 + e^{-x}}$. The asymptote is at $y = 6$, and the y coordinate of the point of inflexion is $6 \div (1 + 1) = 3$.

The only change for the blue graph is the value of k: its equation is $y = \dfrac{6}{1 + e^{-2x}}$. The y-intercept is the same, but we can see that most of the growth is in the range $-2 < x < 2$, rather than $-3 < x < 3$ for the red graph.

For the purple graph, I have used the same values as for the red graph, except for C which is 2. So $y = \dfrac{6}{1 + 2e^{-x}}$. It has a similar shape to the red graph, but the y-intercept is now at $\left(0, \dfrac{6}{1 + 2}\right) = (0, 2)$, thus delaying the main part of the growth.

Given the graph, the values of L and C are easy to compute, but k will either have to be given or calculated given further information, such as the coordinates of a point.

Example: The diagram shows the graph of the logistic function $y = \dfrac{p}{1 + qe^{-kx}}$.

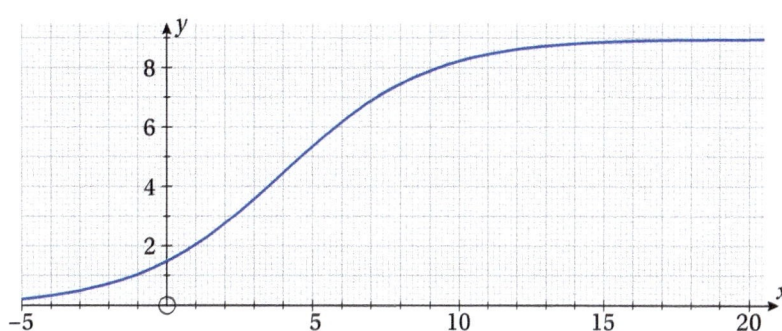

(a)   Use the graph to suggest values for p and q.

(b)   Estimate the y-coordinate when $x = 3$. Hence calculate an estimate for k to an appropriate degree of accuracy, using your answers to (a).

Solution: (a)   The horizontal asymptote appears to be $y = 9$, so $p = 9$. The y-intercept is at $(0, 1.5)$, so $1.5 = \dfrac{9}{1+q} \Rightarrow q = 5$

(b)   When $x = 3$, $y = 3.6$. Thus $3.6 = \dfrac{9}{1 + 5\,e^{-3k}} \Rightarrow k = 0.401$. Since we are reading values from a graph, an estimate for k is 0.4.

Logistic models: A logistic model may be appropriate in situations where there is a restriction on growth.

> The maximum value can be referred to as the *carrying capacity*.

- Rabbits are introduced on a small island. The population grows quickly, but since food resources are limited, it reaches a maximum value.

- The height of a tree after planting. Growth is slow at first, then increases, but the tree eventually reaches a maximum height depending on its species and the available nutrients and water.

- Decrease in the number of cases of a disease on a yearly basis after the start of a vaccination programme.

The table shows the years in which world population reached given values, extrapolated into the future using figures from the results of a study.

Year	1927	1960	1974	1987	1999	2011	2025	2041	2071
Population (billions)	2	3	4	5	6	7	8	9	10

A scattergraph is drawn of the data where t is the number of years after 1900 and p is the population in billions.

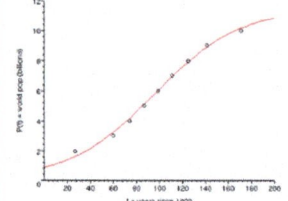

(a)   Explain why a logistic function could be used to model the data.

The logistic function $p = \dfrac{11.5}{1 + 12.8\,e^{-0.0266t}}$ is suggested as a suitable model.

(b)   Sketch the graph of the function for $0 \le t \le 200$, marking any axis intercepts.

(c)   Explain the relevance of the constant 11.5.

(d)   Given that y-coordinate of the point of inflexion occurs midway between the two asymptotes find the value of T at which it occurs.

(e)   In terms of the data, explain the significance of T.

(a)   The scattergraph has the typical S shape of a logistic function, and appears to be approaching a limiting value.

(b)

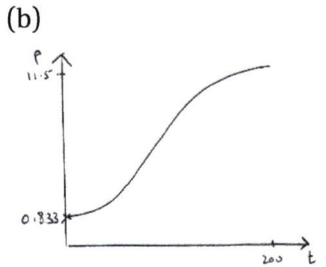

(c)   11.5 indicates that population will reach a maximum value of 11.5 billion

(d)   Asymptotes at $y = 0$ and $y = 11.5$, so y-coordinate of inflexion point $= 11.5 \div 2 = 5.75$.

$5.75 = \dfrac{11.5}{1 + 12.8\,e^{-0.0266t}} \Rightarrow t = 95.84$.

Thus $T = 1996$.

(e)   In 1996 the rate of growth of population began to decrease.

2.11 Using Logarithms in Modelling

Why use a logarithm scale?

There are two situations when a log scale is appropriate:

(a) To manage a wide range of values.

(b) When the emphasis is on the rate of growth rather than the actual values.

As an example of (a), consider the share price of a small mining company which struck gold in 1985:

Year	1980	1981	1982	1983	1984	1985	1986	1987	1988
Price ($)	1.25	1.48	1.56	1.78	2.31	6.80	8.51	10.00	17.45

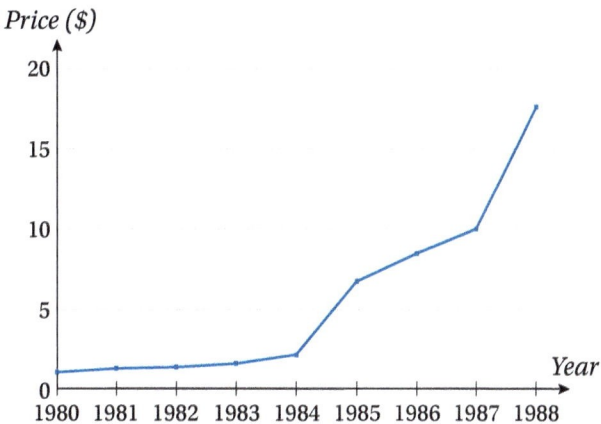

The problem with this graph is that, visually, the variation of the values up to 1984 has been compressed, and appears to be constant, even though the share price has nearly doubled. A log scale removes this problem, although we do lose the equal intervals on the y-axis.

Consider:

$\log_{10} 10 = 1$

$\log_{10} 100 = 2$

$\log_{10} 1000 = 3$

So using a log scale reduces the gaps as the numbers increase.

Year (x)	1980	1981	1982	1983	1984	1985	1986	1987	1988
Price (y)	1.25	1.48	1.56	1.78	2.31	6.80	8.51	10.00	17.45
$\log_{10} y$	0.097	0.170	0.193	0.250	0.364	0.833	0.930	1.00	1.24

Although I have shown the actual price values on the y-axis for clarity, you may find that sometimes the log values are shown instead.

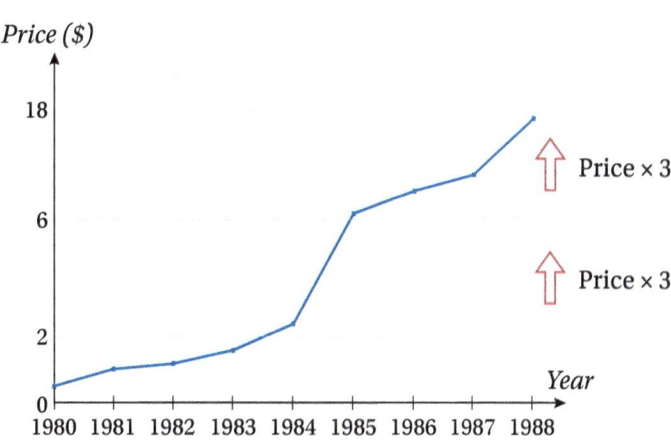

This graph (known as a *semi-log* graph because the x axis is still linear) allows us to read off the lower values as well as the higher. On the y-axis, equal intervals on the scale relate to equal *multipliers* in the data. For example, the price nearly trebled between 1984 and 1985, and then again between 1985 and 1988 – these show as equal intervals, as indicated by the red arrows.

This leads us to the second use of a log scale: when the rate of growth is more important than the values. For example, volume is measured in decibels (dB). A human ear is attuned to increases in volume, rather than actual values – in other words, if a noise doubles in volume, and then doubles in volume again, the ear registers both increases in the same way. For that reason, the dB scale is a log scale. Every increase of 10dB is equivalent to a tenfold increase in sound intensity (roughly corresponding to a doubling in volume).

Linearising data: When a set of data values has been collected and plotted on a graph, the shape of the graph may suggest a possible function as a model. However, only the equations of straight lines can be found easily – there may be several possibilities for curved graphs. For models of the form $f(x) = ax^k$ or $f(x) = ab^x$, we can use the laws of logarithms to transform the functions into equations of straight lines.

I generally use logarithms to the base 10 in which case, once you have found the intercept, calculate a as $10^{\text{intercept}}$. In the second case, find b in the same way.

If you are asked to use ln instead, then use e rather than 10 for the calculations.

$y = ax^k$	$y = ab^x$
$\log y = \log(ax^k)$	$\log y = \log(ab^x)$
$\log y = \log a + \log x^k$	$\log y = \log a + \log b^x$
$\log y = \log a + k \log x$	$\log y = \log a + x\log b$
$\log y = k \log x + \log a$	$\log y = (\log b)x + \log a$
Compare $Y = mX + c$	Compare $Y = mX + c$
Plot $\log y$ against $\log x$ and we will get a straight line where k is the gradient and $\log a$ is the y-intercept.	Plot $\log y$ against x and we will get a straight line where $\log b$ is the gradient and $\log a$ is the y-intercept.

An investigation is carried out into the rate of flow R of water in pipes of differing diameter d for a given water pressure. The following data set was collected:

d (cm)	1	2	3	5	10
R (litres/sec)	0.02	0.31	1.65	12.8	195

It is thought that the relationship between R and d is of the form $R = pd^k$ where p and k are constants.

(a) Explain how the graph of $\log d$ against $\log R$ can help show if this is a good model.

(b) (i) Plot the values of $\log_{10}R$ against $\log_{10}d$.

    (ii) Draw a line of best fit on your graph.

    (iii) Hence estimate the values of p and k.

(c) (i) Using the model with your values of p and k, calculate the value of R for $d = 5$.

    (ii) Find the % error for the value calculated in part (i).

(a) $R = pd^k$

$\log_{10}R = \log_{10}p + k\log_{10}d$

This is of the form $y = mx + c$, where $c = \log_{10}p$ and $m = k$.

Note that we are not using logarithmic scales. We are just plotting values of logarithms on ordinary linear scales.

You should also try to answer this question by plotting the points on your GDC and then using linear regression to find the line of best fit.

(b) (i)

$\log_{10}d$	0	0.30	0.48	0.70	1.00
$\log_{10}R$	-1.70	-0.51	0.22	1.11	2.29

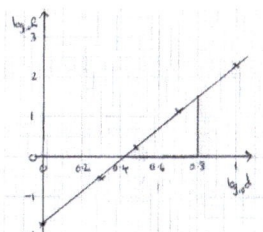

    (ii) Gradient $= 1.5 \div 0.38 = 3.947 \Rightarrow k = 3.95$

Intercept $= -1.7 \Rightarrow p = 10^{-1.7} = 0.02$

(c) (i) $R = 0.02d^{3.95}$

When $d = 5$, $R = 11.53$

    (ii) % error $= \dfrac{12.8 - 11.53}{12.8} \times 100 = 9.9\%$

Functions: Long Answer Questions

Starting on the next page is a selection of Paper 2 style exam questions related to Functions. The answers are given here, but full workings may be found on the Peak Study Resources website.

1. Let $f(x) = \dfrac{20x}{e^{0.3x}}$ where $0 \leq x \leq 20$

   (a) Sketch the graph of f.

   (b) Find the x-coordinate of the maximum point.

   (c) State the range of values of x for which f is an increasing function.

   Let $g(x) = \dfrac{20}{x}$ where $0 \leq x \leq 20$.

   (d) Add the graph of g to your sketch.

   (e) Solve the equation $\dfrac{20x}{e^{0.3x}} = \dfrac{20}{x}$, $0 \leq x \leq 2$, giving your answer to 4SF.

   (f) Find $f(20)$ and $g(20)$ and hence state whether there is a second solution to the equation.

   Answers:

   (b) $x = 3.33$ (a)

   (c) $0 \leq x < 3.33$ (d)

   (e) $x = 1.197$.

   (f) Yes, there is.

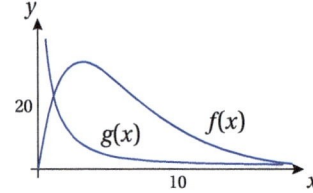

2. The function $f(x)$ is defined by $f(x) = (x - 2)^2 + 1$ for $x \geq k$.

   (a) State the minimum value of k such that the inverse of f is also a function.

   (b) Find $f^{-1}(x)$.

   (c) State the domain and range of f^{-1}.

   (d) The graph of $g(x)$ is obtained by a reflection of the graph of f in the x-axis, then a translation along the vector $\begin{pmatrix} 0 \\ 4 \end{pmatrix}$.

        (i) Find $g(x)$.

        (ii) Solve $f(x) = g(x)$ given the value of k found in part (a).

   (e) Find $h(x)$ in the form $ax + b$, $a > 0$, such that $(f \circ h)(x) = 25x^2 + 20x + 5$

   Answers:

   (a) 2

   (b) $2 + \sqrt{x - 1}$

   (c) $x \geq 1$, $f^{-1}(x) \geq 2$

   (d) (i) $g(x) = 3 - (x - 2)^2$

        (ii) $x = 3$

   (e) $5x + 4$

3.   Let $f(x) = \frac{1}{x} + 3$, $x \neq 0$ and $g(x) = \ln x$, $x > 0$

(a)   The graph of f is reflected in the y–axis and then translated along the vector $\begin{pmatrix} 2 \\ -1 \end{pmatrix}$. What is the equation of the transformed graph?

(b)   Find $h(x) = (f \circ g)(x)$.

(c)   Find the domain and range of h.

(d)   (i)   Show that $h^{-1}(x) = e^{\frac{1}{x-3}}$.

(ii)   Solve the equation $h(x) = h^{-1}(x)$ for $x < 1$.

(iii) Find the x-intercept of $h(x)$.

(iv) Find an expression involving integrals for the area bounded by the x- and y-axes, and the graphs of h and h^{-1}.

(v)   Evaluate the integral in part (iv).

_____

Answers:

(a) $y = 2 - \dfrac{1}{x-2}$

(b) $y = \dfrac{1}{\ln x} + 3$

(c)   Domain is $x > 0$, $x \neq 1$. Range is $h(x) \neq 3$.

(d)   (ii) $x = 0.653$

(iii) $(0.717, 0)$

(iv) Area $= \int_0^{0.653} e^{\frac{1}{x-3}} \, dx + \int_{0.653}^{0.717} \frac{1}{\ln x} + 3 \, dx$

(v) 0.4706

4.   The cost of producing lengths of material is made up of a fixed cost to set up the machine plus an amount for each metre of material produced. The average cost per metre (£C) for producing x metres of material is given by the equation $C = \dfrac{ax + b}{x}$, $x > 0$.

It is given that 20 m of material costs £35 per metre to produce and 40 m of material costs £25.

(a)   Find the value of a and the value of b.

(b)   State the fixed cost for setting up the machine.

The manufacturer plans to sell the material at £20 per metre.

(c)   Find the minimum length that needs to be sold to avoid making a loss.

(d)   Explain why the maximum profit per metre can never be greater than £5.

_____

Answers:

(a) $a = 15$, $b = 400$

(b)   £400

(c)   80m

(d) $x \to \infty$, $C \to 15$ from above.

Cost always > 15, so profit < 5

5. **Part A**

Consider the line L which has equation $2x + 4y - 6 = 0$

P is the point where the line crosses the x-axis.

Q is the point where the line crosses the y-axis.

(a) (i)   Find the coordinates of P.

    (ii) Find the coordinates of Q.

    (iii) Calculate the distance PQ.

    (iv) Calculate the area of triangle OPQ, where O is the origin.

(b) (i)   Calculate the gradient of L.

    (ii) Hence calculate the gradient of a line perpendicular to L.

    (iii) Determine the equation of the line through P which is perpendicular to L.

Part B

The function $E(t) = 0.003\,t^2 + kt + 25$ represents the amount of energy, in kJ, in a battery after t minutes of use.

(a) State the amount of energy held by the battery immediately before it was used.

After 20 minutes of use, the amount of energy held is 13.7 kJ.

(b) Calculate the value of k.

(c) (i)   Calculate the number of minutes before the amount of energy is zero.

    (ii) Hence state the domain of the function E.

_____

Answers:

Part A

    (a) (i)   (3, 0)

        (ii) (0, 1.5)

        (iii) 3.354

        (iv) 2.25

    (b) (i)   –0.5

        (ii) 2

        (iii) $y = 2x - 6$

Part B

    (a) 25 kJ

    (b) $k = -0.625$

    (c) (i)   54

        (ii) $0 \leq t \leq 54$

6. Giulia is doing an experiment in her Physics lesson. In the experiment a weight is hung from a spring and its position recorded as its equilibrium position.

The weight is then raised and released. The weight's displacement from its equilibrium position is y cm and time t is measured in milliseconds from the moment the weight first passes its equilibrium position.

Giulia films the subsequent motion of the weight and from the film finds the lowest point reached by the weight is $y = -2$ when $t = 1.0$ msec, and then passes the equilibrium position once more when $t = 2.0$ msec

The diagram shows images from the film for $t = 0$, 1 and 2.

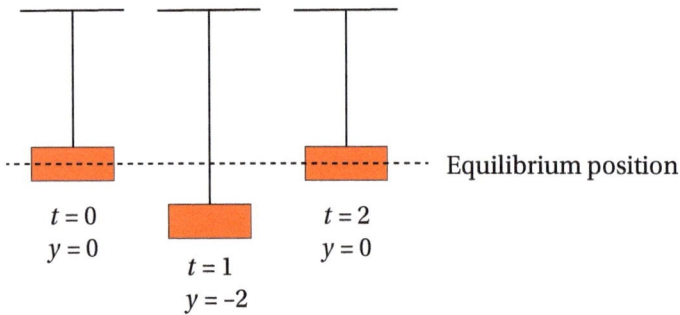

Giulia decides that a suitable model for the motion of the weight will be of the form $y = a \sin bt$.

(a) Write down the value of a.

(b) Find the value of b.

Giulia then decides that a quadratic model, $y = ct^2 + dt + e$, might be more suitable.

(c) Explain why $e = 0$.

(d) Find the values of c and d.

(e) Use both models to predict the displacement of the weight from its equilibrium position when $t = 3$.

(f) Hence explain which is the more appropriate model for the given situation.

_____

Answers:

(a) $a = -2$

(b) 90

(c) Because when $t = 0$, $y = 0$

(d) $c = 2$, $d = -4$

(e) Model 1 gives 2 cm, model 2 gives 6 cm.

(f) Model 1 is better since in model 2, y will always increase with t for $t > 1$

7. The temperature of a hot drink t minutes after pouring is thought to be given by the function $T = ab^t + c$.

    To test this theory the temperature of the drink is taken at two minute intervals and the results recorded:

t (min)	2	4	6	8	10
T (°C)	62	44	36	30	24

    It is given that the temperature of the room is 22°C.

    (a) Write down the value of c.

    (b) (i)   Draw a graph of t against $\ln(T - c)$.

         (ii) Explain why your graph suggests that the model is valid.

         (iii) Evaluate a statistic to justify your answer to (b)(ii).

    (c) (i)   Write down the equation of the line of best fit.

         (ii) Hence find the values of a and b.

    _____

Answers:

(a) $c = 22$

(b) (i)

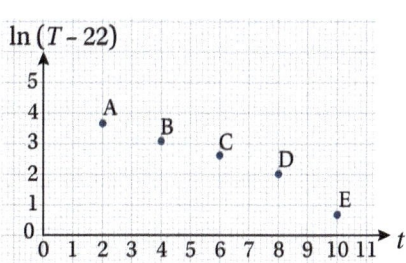

     (ii) The points show a linear model

     (iii) $r = -0.971$. Strong negative correlation.

(c) (i) $\ln(T - 22) = 4.54 - 0.350t$

     (ii) $b = -0.350$, $a = 93.7$

Chapter 3: GEOMETRY AND TRIGONOMETRY

3.1 Solution of Triangles

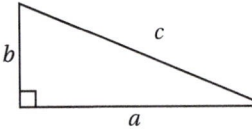

Right-angled triangles: This section is a reminder of how to deal with the sides and angles of a right-angled triangle. The following section deals with non right-angled triangles.

Pythagoras' Theorem: If you know two sides of a right-angled triangle, you can calculate the third using Pythagoras' Theorem. This states that the square of the hypotenuse (the longest side) equals the sum of the squares of the two shorter sides. As applied to the diagram, $c^2 = a^2 + b^2$. You must remember to subtract if you already have the hypotenuse (it's always opposite the right angle) and want to calculate one of the other sides. For example, $b^2 = c^2 - a^2$.

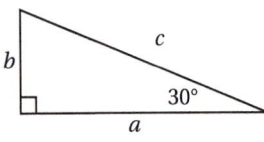

Trigonometry: There is no mystery to sin, cos and tan. They simply represent the ratios of pairs of sides for a triangle with given angles. For example, suppose the smallest angle in the triangle is 30°. Whatever the size of the triangle, b turns out to be half of c. The ratio of b to c is called the sine (sin for short), so $\sin 30° = 0.5$. The ratio of a to c is called the cosine (cos), and b to a is the tangent (tan). If you use the following procedure *in all cases* then every question can be worked out in the same way, and you should always get the right answer.

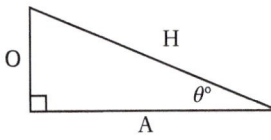

1. Label the three sides of the triangle with H (for hypotenuse, the side opposite the right angle), O (for opposite, the side opposite the angle you are dealing with) and A (for adjacent, the side next to the angle).

2. For the two sides you are dealing with, write down the word sin, cos or tan according to the mnemonic SOH/CAH/TOA.

3. Now write down the angle (which may be unknown) followed by an equals sign.

4. On the right hand side of the equals sign, you will write down a fraction (O over H, A over H, or O over A) which will either involve two known sides, or one known and one unknown side.

5. You will now have an equation to solve. The three examples below show how to do this.

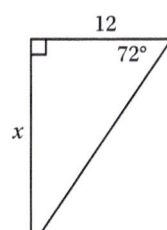

Example 1: Find x.

x is O, 12 is A, so we use tan.

Write down tan, then the angle, then =, then the fraction O/A.

To solve this equation, just multiply through by 12.

$$\tan 72° = \frac{x}{12}$$

$$12 \times \tan 72° = x$$

$$x = 36.9°$$

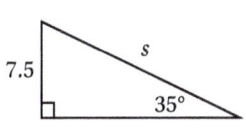

Example 2: Find s.

s is H, 7.5 is O, so we use sin.

Write down sin, then the angle, then =, then the fraction O/H.

The unknown is now on the bottom of the fraction, so we must cross-multiply to find s.

$$\sin 35° = \frac{7.5}{s}$$

$$s = \frac{7.5}{\sin 35°}$$

$$s = 13.1$$

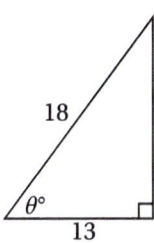

Example 3: Find the angle $\theta°$.

13 is A, 18 is H, so we use cos.

Write down cos, then the angle, then =, then the fraction A/H.

Calculate the value of the fraction, then use the $\cos^{-1}$ function to find out the angle ($\cos^{-1}$ means "find the angle whose cosine is...)

$$\cos \theta = \frac{13}{18}$$

$$\cos \theta = 0.7222$$

$$\theta = \cos^{-1} 0.7222$$

$$\theta = 43.8°$$

Having worked out $\cos \theta$ leave the answer on the display. Then work out the angle using $\cos^{-1}$ANS. This ensures full accuracy.

Sine and Cosine Rules: For triangles which are ***not*** right-angled we use the sine and cosine rules. The triangle shown has the conventional notation of small letters for the lengths of sides and capital letters for the angles opposite. To find lengths and angles, use:

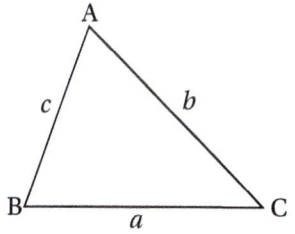

- The sine rule if 2 sides and 2 angles are involved, unless one of the angles is between the two sides

- The cosine rule if 3 sides and 1 angle are involved

SINE RULE
$\dfrac{a}{\sin A} = \dfrac{b}{\sin B} = \dfrac{c}{\sin C}$

COSINE RULE
$c^2 = a^2 + b^2 - 2ab \cos C$ *(for a side)*
$\cos C = \dfrac{a^2 + b^2 - c^2}{2ab}$ *(for an angle)*

Don't be put off by the letters. Basically, the sine rule says the ratio of side/sine is the same for each pair of sides and angles. And in both versions of the cosine rule, ensure that the side c matches the angle C.

In triangle ABC, angle B = 43°, AC = 6.8 cm and AB = 4.3 cm. Find the size of angle A giving your answer to the nearest degree.

$$\frac{\sin C}{4.3} = \frac{\sin 43}{6.8}$$

$$\sin C = \frac{4.3 \sin 43}{6.8} = 0.4313$$

$$\therefore C = 25.55°$$

$$\text{So } A = 180 - (43 + 25.55) = 111.45$$

$A = 111°$ to the nearest degree.

It's very useful to draw a quick sketch to see how to proceed.

We know 2 sides and 1 angle, and we want another angle, so we use the sine rule (which I've inverted to make the calculation easier – always start by writing the thing you want to work out).

In this case we can only work out C from the sine rule, but then we can use the sum of the angles to find A. Note that you should always work to more figures than you need.

Example: A triangle has sides 4, $\sqrt{48}$ and 8. Calculate the size of the angle opposite the side with length $\sqrt{48}$.

Make life easy for yourself by remembering that you don't need a calculator to find the square of a square root!

Solution: This time we need the cosine rule in its second form, making sure that the side labelled c in the formula is opposite the required angle.

$$\cos C = \frac{4^2 + 8^2 - (\sqrt{48})^2}{2 \times 4 \times 8}. \text{ Check that this gives } C = 60°.$$

Ambiguous case using the sine rule: Suppose we are given a triangle where AC = 8, BC = 5, and angle A = 30°. The diagram shows that with this information there are two possible triangles which can be drawn, and hence two possible values for angle B – this is known as the *ambiguous case*. Having found one answer, the other can be found by subtracting from 180°. In this case, B_1 is 126.9° and B_2 is 53.1°.

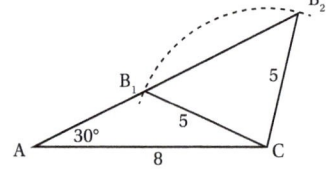

Note that there won't always be two answers – it depends on the given numbers.

Area of a non-right angled triangle: If you know two sides of a triangle, and the size of the angle between the two sides, then the area of the triangle can be found using: Area = $\frac{1}{2}ab \sin C$.

But don't forget the alternative formula $A = \frac{1}{2} \times$ base $\times$ height which is useful if two sides are perpendicular.

Example: A triangle has sides 5, 7, and 8. Find the size of the smallest angle and the area of the triangle.

Solution: The smallest angle is opposite the shortest side. Using the cosine rule we get $\cos x = \frac{7^2 + 8^2 - 5^2}{2 \times 7 \times 8} = 0.786$.

This gives an angle of 38.2°. The sides either side of this angle are 7 and 8, so the area is $\frac{1}{2} \times 7 \times 8 \times \sin 38.2° = 17.3$

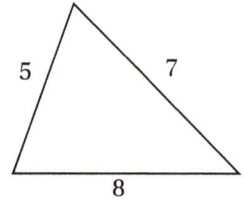

Bearings: One of the practical applications of non-right angled trigonometry is the calculation of distances and angles for moving ships and planes. Their direction of travel is based on compass directions, called *bearings*. A bearing is an angle measured around clockwise from North. Always draw in North lines on your diagrams before marking in bearings.

If a question involves bearings between places, check whether you are dealing with the bearing of A *from* B or the bearing from A *to* B, which is the other way round. Use arrows to show in which direction to take the bearing, and put the North line at the *start* of the arrow.

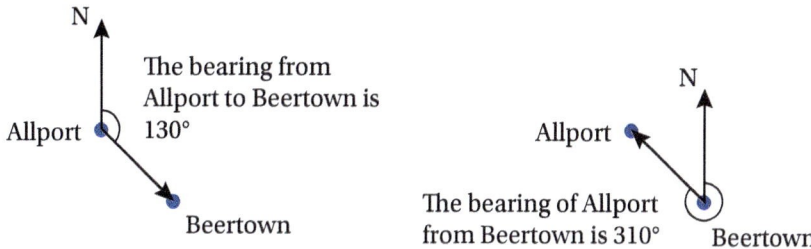

If you find calculations with bearings a bit confusing, I suggest you work through the following question with me. It's important that you work with a large, clear sketch. Also, at the risk of repeating myself, note that I have worked to 4SF in order to give accurate answers to 3SF.

A ship sails from port P and travels due South to port Q. From port Q it sails on a bearing of 065° and travels for 45 km to a point R, which is due East of P.

(a) (i)   Draw and label clearly a diagram to show P, Q and R.

(ii)   Calculate the distance from port P to point R.

In questions like this the diagram is an important tool, so make it large. Angles do not have to be accurate or lengths drawn to scale, but make them look approximately right.

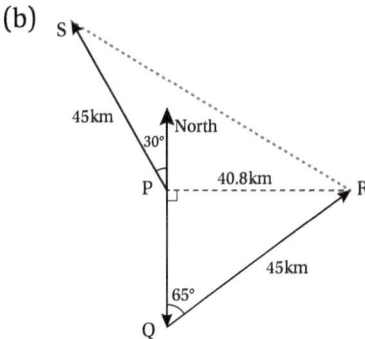

(ii) $\sin 65° = \dfrac{PR}{45} \Rightarrow PR = 45\sin 65° = 40.8$

The distance from P to R is 40.8 km

A second ship also sails from port P for 45 km to a point S, but on a bearing of 330°.

(b)   Complete your diagram in part (a) to show point S.

(c)   Calculate the distance from R to S (shown with a grey dotted line) and the angle PRS.

(b)

Rather than putting in 330°, the more useful 30° has been shown instead. The 40.8 has also been put in: always keep your diagrams up-to-date with new information.

To calculate RS, we use triangle PRS which is not right angled. We already know two sides and one angle (SPR = 30 + 90 = 120°), so we use the cosine rule:

73

(c) $RS^2 = 45^2 + 40.783^2 - 2 \times 45 \times 40.783 \times \cos 120°$

$RS = \sqrt{5523.5} = 74.3\,km$

(Check: RS < RP + PS. Looks OK)

Now we need to calculate angle PRS. We know one angle and two sides so we use the sine rule.

$\dfrac{\sin PRS}{45} = \dfrac{\sin 120}{74.321} \Rightarrow \sin PRS = \dfrac{45\sin 120}{74.321} = 0.5244$

So angle $PRS = \sin^{-1}(0.5244) = 31.6°$

(d) What is the bearing of S from R?

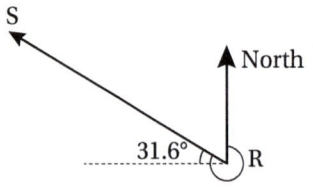

The diagram shows the arrow representing S from R, and a new North line inserted. The required bearing has also been put in. How big is this angle? From North round to West is 270°, and then we need another 31.6.

(d) The bearing of S from R is $270 + 31.6 = 301.6°$

Solution of Triangles: Practice Exercise

Answers

1. $b = 5.32\,cm$, $c = 7.20\,cm$
2. $p = 4.31\,cm$, $Q = 79.9°$
3. $80.3\,m$, $2.87°$
4. $10.9\,km$, $059.4°$
5. $14.5\,cm$
6. $A = 36.9°$, $B = 90°$, $C = 53.1°$
7. $9.64\,cm$

1. Triangle ABC has A = 66°, B = 44°, a = 7cm. Find b and c.

2. Triangle PQR has P = 45°, r = 5cm and q = 6cm. Find p and Q.

3. a) An observer on a ship which is 2.3km from the coast measures the angle of elevation of a cliff as 2°. Find the height of the cliff in metres.

   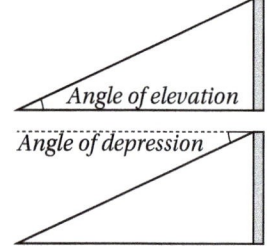

   b) A 35m tower stands on the top of the cliff. Find the angle of depression for an observer looking at the ship from the top of the tower.

4. A ship S is 7km away from a lighthouse L on a bearing of 080° and a ship T is 5km away from L on a bearing of 210°. Find the distance and bearing of S from T.

5. A rhombus has sides of length 8cm and angles of 50° and 130°. Find the length of the longer diagonal of the rhombus.

6. A triangle ABC has area 24cm² and sides a = 6cm, b = 10cm. Find all the angles in the triangle.

7. Calculate WX, given YZ = 15cm.

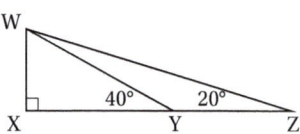

3.2 3-D Geometry

You have to be able to combine the rules of 3-D coordinates with the trigonometry of both right angled and non-right angled triangles in order to find the sides and angles of cuboids, prisms and pyramids.

Cuboid: A cuboid is the 3-D equivalent of a rectangle. It has 12 edges, 4 each in three different dimensions (length, width and height). Commonly asked questions involve the lengths of diagonals (both of sides and also from one corner to the opposite corner) and angles between various lines. The points used will be the vertices (corners) and the midpoints of sides.

Volume of a cuboid: $V = $ length $\times$ width $\times$ height

Pyramid: You only have to concern yourself with a "right" pyramid – ie where the apex is directly above the centre of the base, which is itself a square. Pyramid questions almost invariably use the midpoints of sides, and it should be noted that a line drawn from the midpoint of one of the base edges to the apex is at right angles to the base edge.

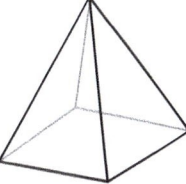

Volume of a pyramid: $V = \frac{1}{3} \times$ base area $\times$ height

Prism: A prism is any 3-D shape with the same cross-section throughout its length. Very often this cross-section is a triangle, but it does not have to be.

Volume of a prism: $V = $ area of cross section $\times$ length

Angle between a line and a plane: A plane is a flat surface, so each of the faces of the 3-D shapes illustrated is a plane.

The angle between a line and a plane is the angle between the line and its *projection* on the plane: think of the projection as part of a "shadow line" on the plane.

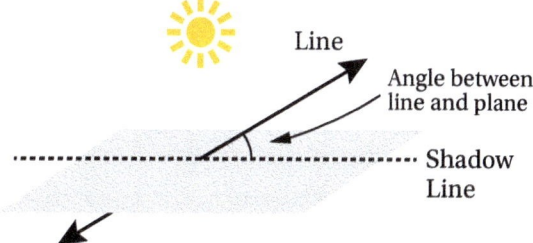

Calculating lengths and angles in 3 dimensions: In every case, you convert the question into an appropriate 2-D question, usually by identifying a right-angled triangle containing the length/angle you have to work out, drawing it as it really looks, then using trigonometry and Pythagoras as usual.

For example, the base edges of a pyramid are 10 cm and the slant edges are 12 cm. M is the midpoint of side PQ and X is the centre point of the base. Find length AM and angle AMX to the nearest degree.

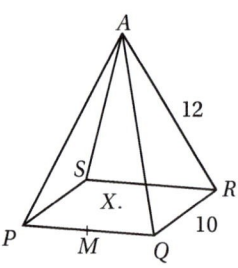

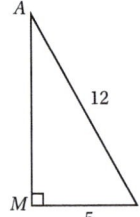

No need to calculate $\sqrt{119}$ since we are using it in the next stage.

We can find AM using triangle AMQ. MQ = 5 (half of the base length)

So $AM^2 = 12^2 - 5^2 \Rightarrow AM = \sqrt{119}$.

Now we can draw triangle AMX because we know MX = 5 and $AM = \sqrt{119}$.

We can see from the diagram that $\cos M = \dfrac{5}{\sqrt{119}}$ giving M = 62.7°, or 63° to the nearest degree.

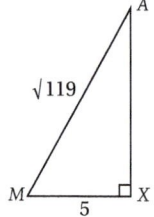

A prism ABCDEF is formed with an isosceles triangle cross section and a rectangular base. AE = 6.4 cm, AB = 5 cm, BC = 11 cm. M is the midpoint of AB.

  (a) Find the length of EM.

  (b) Hence find

    (i)   The area of triangle AEB

    (ii) The volume of the prism

  (c) Find the length of MC, and hence the angle EC makes with base ABCD.

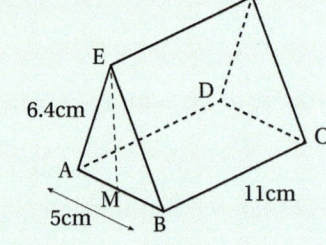

(a) $EM^2 + 2.5^2 = 6.4^2$

      $EM^2 = 34.71$

      $EM = 5.89$ cm

(b) (i)   Area = $\frac{1}{2}$ × base × height

             = $\frac{1}{2}$ × 5 × 5.89 = 14.7 cm²

      (ii) Volume = 14.7 × 11 = 162 cm³

(c) $MC^2 = 2.5^2 + 11^2$

      MC = 11.3 cm

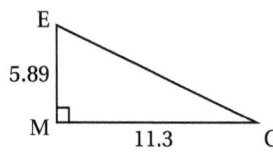

      $\tan(ECM) = \dfrac{5.89}{11.3}$

      Angle EC makes with base is 27.6°

(a) Wherever possible, try to find right-angled triangles which will lead to the solution. In this case we can use EAM, knowing that AM is one half of AB.

(b) Use standard formula for the area of a triangle and the volume of a prism.

(c) Once again we find in EMC a right-angled triangle which will give us MC. It also gives us the angle EC makes with the base.

As before, I have given answers to 3SF, but worked to 4SF.

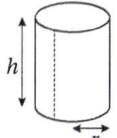

3.3 Cylinder, Sphere and Cone

Curved Surface Area: Whilst the concept of the volume of shapes with curved edges and faces may not be too difficult to appreciate, the area of such curved faces may be problematic. The way to think of such faces is to imagine them to be made from separate pieces of paper; the area of a curved surface is the same as the area of the paper it is made from when it is flattened out. For example, unroll the curved surface of a cylinder and you get a rectangle whose height is the height of the cylinder and whose length is the circumference of the cylinder.

Cut along the dotted line and open up to get...

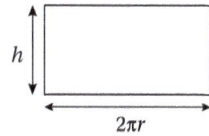

You need to be able to use the relevant area and volume formulae for cylinders, cones and spheres. You will find them all in your list of formulae. They are:

Cylinder: Curved surface area $= 2\pi rh$

Volume $= \pi r^2 h$

Cone: Curved surface area $= \pi r l$ (l is the slant height)

Volume $= \frac{1}{3}\pi r^2 h$

Sphere: Surface area $= 4\pi r^2$

Volume $= \frac{4}{3}\pi r^3$

Read each question very carefully to see *exactly* what you are being asked to find. For example, a cylinder may be completely closed in which case the total surface area is the curved surface area plus the areas of the two ends, which are both circles. Or it may be open at one end, so just add one circle.

The volume of a hemisphere is half that of a sphere, but its total surface area will be half the curved surface of a sphere plus the circle which forms its base. Its formula will be $\frac{1}{2} \times 4\pi r^2 + \pi r^2 = 3\pi r^2$.

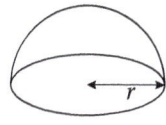

Example: Both a cone and a hemisphere have base diameter of 18 cm. If the height of the cone is 10 cm, show that the ratio of the volume of the cone to that of the hemisphere is 5:9.

Solution: The volume of the cone is $\frac{1}{3}\pi r^2 h = \frac{1}{3}\pi \times 9^2 \times 10 = 270\pi$. The volume of the hemisphere is $\frac{1}{2} \times \frac{4}{3}\pi r^3 = \frac{2}{3}\pi \times 9^3 = 486\pi$. We may as well leave π in the answers since we are finding the ratio.

There's an old trap in this question – all the formulae use the radius, but we have been given the diameter.

So, volume of cone:volume of hemisphere = $270\pi : 486\pi = 5:9$

Try the following example for yourself.

Example: In the South of England, special brick constructions called "Oast Houses" were built to dry the hops used in making beer. The diagram models an Oast House as a cone on top of a cylinder. Both have base diameter 8 m; the cylinder has a height of 6 m, and the building is 17.4 m high overall.

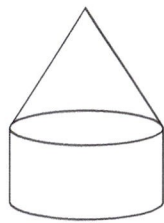

Find the total surface area of the exposed faces of the Oast House, and the total volume.

You will need the slant height of the cone to find the curved surface area – use Pythagoras' Theorem. Note too that the base of the cone will not form part of the total surface area since it is not an external surface.

Answers: Area = 303 m², Volume = 493 m³.

Full working can be found on the website

A cylindrical tube of length 40 cm and radius 3.2 cm contains 6 tennis balls each with a radius of 3.14 cm.

    (a)   Giving answers to 3SF, find:

        (i)   The volume of the tube

        (ii)   The volume of one tennis ball

    (b)   Hence find the volume of empty space in the tube.

(a) (i) $V = \pi r^2 l = \pi \times 3.2^2 \times 40 = 1286.7...$

        Volume of tube = 1290 cm³

    (ii) $V = \frac{4}{3}\pi r^3 = \frac{4}{3}\pi \times 3.14^3 = 129.68...$

        Volume of ball = 130 cm³

(b) Volume of 6 tennis balls

              = 129.68 × 6 = 778.1 cm³

    Empty space   = 1286.7 – 778.1 = 508.6

               = 509 cm³ to 3SF

If you're going to use a formula, write it down. This shows the examiner what you are using, and helps you to substitute values correctly.

This question is yet another good example of the importance of working to a higher accuracy than the required answer. If the first two rounded answers were used to obtain the final answer, the result would have been

1290 – 6 × 130 = 510.

Geometry: Practice Exercise

Answers

1. 5.44 km, 272.2°

2. 10 cm, 13 cm, 67.4°, 12.4 cm, 76.0°, 5 cm, 40.6°

3. Volume of cone = 25.1 cm³

   Volume of each sphere = 0.524 cm³

   47 spheres, with a bit left over!

1.   Three boats P, Q and R are at anchor in a bay. The bearing of P from R is 046°, and of Q from P is 125°. The distance of R from P is 3 km, and of P from Q is 4 km.

      (a)   Draw a clear diagram showing all the information.

      (b)   Calculate:

          (i)   the distance of R from Q;

          (ii)   the bearing of R from Q.

2.   The diagram shows a right pyramid with:

      PQ = 8 cm, QR = 6 cm, VW = 12 cm.

      W is at the centre of the base, M and N are the midpoints of PQ and QR. Calculate:

      (a)   PR

      (b)   PV

      (c)   Angle VPW

      (d)   VM

      (e)   The angle VM makes with the base

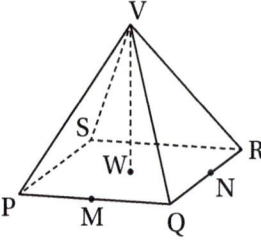

(f)   MN

(g)   Angle MSN (Draw triangle MSN first)

3.   A solid cone of height 6 cm and radius 2 cm is melted down to be made into small spheres. If each sphere has a radius of 0.5 cm, how many can be made?

3.4 Radian Measure

Radians: Radians are an alternative to degrees when measuring the size of angles. Although it is easier to *think* in degrees, radians are often used with trigonometric functions and *must* be used when differentiating or integrating them.

The conversion is π radians = 180°. (An angle is assumed to be in radians unless the degrees symbol is given).

It is worth memorising some key angles in radians (see table in the notes box). π appears in many angles when expressed in radians (because of the conversion) but it does not have to. For example, 45° = 0.785 radians, but this is not an *exact* conversion, unlike $\frac{\pi}{4}$.

There are two circle formulae which are used when a sector angle is expressed in radians. If the angle is q and the radius of the circle is r:

- Arc length of sector = $r\theta$

- Area of sector = $\frac{1}{2}r^2\theta$

30° = π/6
45° = π/4
60° = π/3
90° = π/2
120° = 2π/3
180° = π
270° = 3π/2
360° = 2π

Sectors of Circles: The green shaded area on the diagram is called a *sector*. There are two formulae which are used when a sector angle is expressed in radians. If the angle is θ and the radius of the circle is r:

- Arc length of sector = $r\theta$

- Area of sector = $\frac{1}{2}r^2\theta$

Example: The shaded area ABCD is formed by enclosing the area between the sectors of two circles with radii 3 cm and 4.5 cm. The angle O at the centre of the circles is 135°.

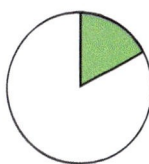

(a)   Express the angle O in radians as a multiple of π.

(b)   Find the area of the shaded region, answer to 1 DP.

(c)   The perimeter of the shaded region, answer to 1 DP.

Solution: (a) $135° = \frac{3\pi}{4}$ radians

(b)   To calculate the area, we subtract the area of sector OBC from the area of sector OAD.

Area of sector OAD = $\frac{1}{2} \times \frac{3\pi}{4} \times 4.5^2 = 23.856$ cm²

Area of sector OBC = $\frac{1}{2} \times \frac{3\pi}{4} \times 3^2 = 10.603$ cm²

Shaded area = 23.856 – 10.603 = 13.3 cm² to 1DP.

(c)   The perimeter = AB + CD + arc BC + arc AD

$$AB = CD = 4.5 - 3 = 1.5$$

$$\text{arc BC} = \frac{3\pi}{4} \times 3 = 7.069$$

$$\text{arc AD} = \frac{3\pi}{4} \times 4.5 = 10.603$$

Perimeter of ABCD = 1.5 + 1.5 + 7.069 + 10.603 = 20.7 cm

3.5 Voronoi Diagrams

Be aware of the notation: [AB] denotes a line segment (ie the part of the line between two points), whereas (MP) denotes a line which extends beyond the two points which define it.

Perpendicular bisectors: Given two points A and B, the *perpendicular bisector* of line [AB] is another line at right-angles to [AB] passing through its midpoint. All points on the perpendicular bisector are equidistant from both A and B.

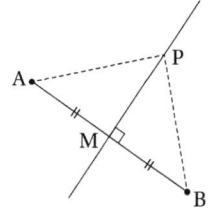

In the diagram, M is the midpoint of [AB], and (MP) is the perpendicular bisector. For any point P on the bisector, AP = BP.

You will not be required to construct the perpendicular bisector, but you need to be able to find its equation either given the points A and B, or the equation of the line segment [AB] and the midpoint M.

You may want to refresh your memory about the various formulae for straight line geometry – see page 44.

Finding the bisector given two points: Suppose the two points are A(5, 8) and B(9, 2):

- Midpoint of [AB] is (7, 5)
- Gradient of (AB) is $-\frac{6}{4}$ which simplifies to $-\frac{3}{2}$
- Gradient perpendicular to (AB) is $\frac{2}{3}$
- Therefore equation of perpendicular bisector is $y - 5 = \frac{2}{3}(x - 7)$
- Alternative forms: $2x - 3y + 1 = 0$ or $y = \frac{2}{3}x + \frac{1}{3}$

Finding the bisector given the equation of the line segment and its midpoint: Suppose the line joining A to B has equation $2x - 4y = 6$ and the midpoint of [AB] is (1, –1):

- Rewrite the equation of (AB) as $y = \frac{1}{2}x - \frac{3}{2}$
- So gradient of (AB) is $\frac{1}{2}$
- Gradient perpendicular to (AB) is –2
- Therefore equation of perpendicular bisector is $y + 1 = -2(x - 1)$
- Alternative forms: $2x + y - 1 = 0$ or $y = -2x + 1$

Perpendicular bisectors: Quick Practice

Answers

(a) $x + 5y = 30$

(b) (30, 0)

(c) PS = ST = 26

(a)   Find the equation of the perpendicular bisector l of the line joining the points S(4, 0) and T(6, 10)

(b)   Find the point P where l intersects the x-axis.

(c)   Show that P is equidistant from S and T.

I've rewritten the definition several times – I hope it makes sense! Much easier to look at the diagram.

What are Voronoi Diagrams?   Take a set of points, called *sites*, in a plane; divide the plane into regions each containing one site such that all the points in that region are closer to the site than to any other site.

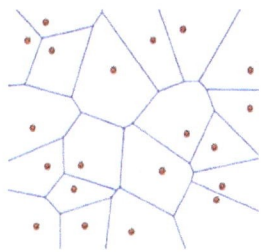

In the diagram, each dot is a site. Look at the region, or *cell*, around any particular site; you can see that all the points in the cell are closer to that site than to any other. The boundaries of each cell, known as *edges*, must be lines which are equidistant from two neighbouring sites – in other words, they are perpendicular bisectors. It also follows that the points where edges meet, known as *vertices*, are always equidistant from three neighbouring sites.

Also note that:

- Every vertex is the intersection of three edges.
- A circle centred on a vertex can be drawn such that it exactly intersects the three nearest sites.
- Every region will be a convex polygon, except …
- … there will always be open regions around the edge of the diagram.

Voronoi diagrams have a wide range of real world applications. One of the most famous (although it wouldn't have been known as a Voronoi diagram then) is John Snow's map relating to a cholera outbreak in London in 1854. On it, he marked the locations of all the known cholera cases and also the taps or pumps which were the local water supplies. He postulated that the cholera was due to contaminated water in those pumps closest to the infected inhabitants; until then, it was though that cholera was caused by "bad air."

The diagram shows a representation of Snow's map with Voronoi cells superimposed. The triangles are the water pumps, and each is a Voronoi site.

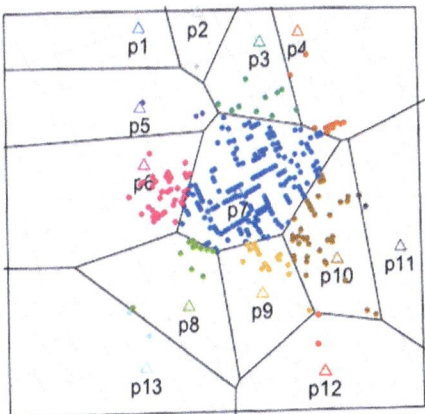

Other applications include:

- Finding the nearest airfield for an emergency landing
- Finding the best place to locate a new wifi point in a wireless network
- Planning a location for a new school
- Many other applications in Biology, Meteorology, Metallurgy, Geography, Astronomy, Medicine…

Nearest neighbour interpolation: The points in a Voronoi cell are nearer to the site in that cell than to any other site. If the site has a value eg: average daytime temperature, then it is assumed that all the points in that cell have the same value. Clearly this isn't necessarily true, and the value of points may in reality change along a gradient towards the next site, but you can always make this simplification in this course.

"Toxic waste dump" problem: Also known as the "largest empty circle" problem. Since any Voronoi vertex is equidistant from the three neighbouring sites, a circle centred on the vertex with a radius just less than the distance to any of the sites will be empty – it cannot have any other sites inside. If the sites represent city centres, then a good place to position a new waste site is at the vertex which is the centre of the largest empty circle.

Many problems can be solved by finding the largest empty circle, such as in the following example.

Example: The Voronoi diagram shows the locations of some of the stores in a large retail chain. The stores are labelled A–G, and form the Voronoi sites.

It is decided to build a new store at either location S_1 or S_2.

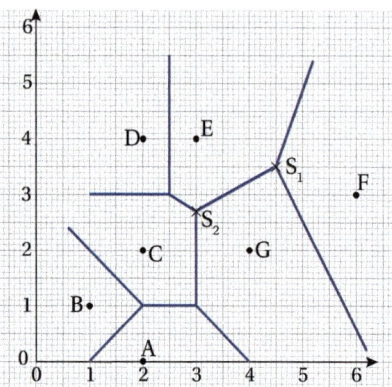

(a) Write down the labels of the three sites closest to S_1.

(b) Hence calculate the radius of the empty circle centred on S_1.

(c) Calculate the radius of the empty circle centred on S_2, given that $S_2 = (3, 2.75)$.

(d) State, with a reason, where the new store should be built.

(e) Calculate the ratio of the areas of the two circles.

Solution: (a) E, F, G

(b) $S_1 = (4.5, 3.5)$ $F = (6, 3)$ (*or use E or G*)

$S_1F = \sqrt{(3 - 3.5)^2 + (6 - 4.5)^2} = \sqrt{2.5} = 1.581$

(c) Using point G:

$S_2G = \sqrt{(4 - 3)^2 + (2 - 2.75)^2} = \sqrt{1.5625} = 1.25$

(d) Build the store at S_1 because it is the centre of the larger empty circle **or** because it will be closer to a larger population than S_2.

(e) We already have the squares of the radii in parts (b) and (c), so the ratio of the areas is $1.25\pi : 1.581\pi = 1 : 1.6$

Answers

1.18 to 1.803

Both 1.581

CG: $x = 3$

EG: $y = 0.5x + 1.25$

$S_2 = (3, 2.75)$

You could also be asked other questions about distance. For example, prove that the location (3.5, 1) is closer to G that it is to A; or prove that (5.5, 1.5) is equidistant from F and G.

Questions may ask you to find perpendicular bisectors. For example, find the equations of the perpendicular bisectors of [CG] and [EG], and hence calculate the coordinates of S_2.

3.6 Matrices and Transformations

The basics of matrices are covered in chapter 1.11 on page 20:

You only need to perform matrix transformations in two dimensions.

Matrix transformations: If we take the position vector of point (p, q) and translate it along the vector $\begin{pmatrix} 3 \\ 4 \end{pmatrix}$ then the transformation can be represented as the matrix addition $\begin{pmatrix} p' \\ q' \end{pmatrix} = \begin{pmatrix} p \\ q \end{pmatrix} + \begin{pmatrix} 3 \\ 4 \end{pmatrix}$, where the image point is (p', q'). Less obviously, the standard geometric transformations of reflections, stretches and rotations can all be represented by matrix multiplication (as long as the origin remains invariant).

Consider the rectangle ABCD where A= (0, 0), B = (1, 0), C = (1, 3), D = (0, 3). We can multiply the position vector of each point by the matrix $\begin{pmatrix} 0 & -1 \\ 1 & 0 \end{pmatrix}$ as follows:

$$\begin{matrix} A & B & C & D \end{matrix} \quad \begin{matrix} A' & B' & C' & D' \end{matrix}$$
$$\begin{pmatrix} 0 & -1 \\ 1 & 0 \end{pmatrix}\begin{pmatrix} 0 & 1 & 1 & 0 \\ 0 & 0 & 3 & 3 \end{pmatrix} = \begin{pmatrix} 0 & 0 & -3 & -3 \\ 0 & 1 & 1 & 0 \end{pmatrix}$$

and plotting the set of image points shows that each of the points, and therefore the rectangle as a whole, has been rotated by +90° around the origin. We can see why this is so by transforming the general point (x, y).

$$\begin{pmatrix} x' \\ y' \end{pmatrix} = \begin{pmatrix} 0 & -1 \\ 1 & 0 \end{pmatrix}\begin{pmatrix} x \\ y \end{pmatrix} = \begin{pmatrix} -y \\ x \end{pmatrix}$$

and the diagram below then shows that the point has been rotated by +90°.

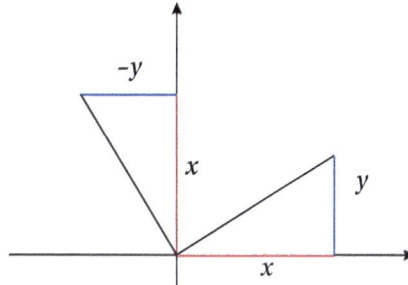

We can also identify which transformation is performed by a particular matrix operation by considering the unit square, which consists of the points (0, 0), (1, 0), (0, 1) and (1, 1). For a matrix $\begin{pmatrix} a & b \\ c & d \end{pmatrix}$, (0, 0) remains unchanged, $(1, 0) \rightarrow (a, c)$ and $(0, 1) \rightarrow (b, d)$.

So:

This works both ways. Consideration of the unit square allows us to identify the transformation given the matrix, and also the matrix given the transformation.

These matrix transformations are all in your formula book.

$\begin{pmatrix} -1 & 0 \\ 0 & 1 \end{pmatrix} \equiv$ Reflection in the y-axis

$\begin{pmatrix} 0 & 1 \\ 1 & 0 \end{pmatrix} \equiv$ Reflection in the line $y = x$

$\begin{pmatrix} a & 0 \\ 0 & 1 \end{pmatrix} \equiv$ Stretch $\times\, a$ parallel to the x-axis

$\begin{pmatrix} 1 & 0 \\ 0 & a \end{pmatrix} \equiv$ Stretch $\times\, a$ parallel to the y-axis

$\begin{pmatrix} a & 0 \\ 0 & a \end{pmatrix} \equiv$ Enlargement $\times\, a$ centre (0, 0)

$\begin{pmatrix} 0 & -1 \\ 1 & 0 \end{pmatrix} \equiv$ Rotation of +90° centre (0, 0)

$\begin{pmatrix} 0 & 1 \\ -1 & 0 \end{pmatrix} \equiv$ Rotation of -90° centre (0, 0)

$\begin{pmatrix} -1 & 0 \\ 0 & -1 \end{pmatrix} \equiv$ Rotation of 180° centre (0, 0)

$\begin{pmatrix} \cos\theta & -\sin\theta \\ \sin\theta & \cos\theta \end{pmatrix} \equiv$ Rotation of $\theta°$ centre (0, 0)

$\begin{pmatrix} 1 & 0 \\ 0 & -1 \end{pmatrix} \equiv$ Reflection in the x-axis

Determinant: The determinant of the matrix also represents the area scale factor of the transformation. For example, the determinant of the stretching matrix $M = \begin{pmatrix} 3 & 0 \\ 0 & 1 \end{pmatrix}$ is 3, and any shape transformed by M will have its area multiplied by 3.

Composite transformations: The matrix multiplication $C = AB$ is equivalent to: "Transformation C = transformation B followed by transformation A." For example, if we look at $\begin{pmatrix} 0 & -1 \\ 1 & 0 \end{pmatrix}\begin{pmatrix} 0 & 1 \\ 1 & 0 \end{pmatrix} = \begin{pmatrix} -1 & 0 \\ 0 & 1 \end{pmatrix}$. As transformations, this becomes: Reflection in $y = x$ followed by rotation of +90° around the origin = reflection in the y-axis.

It is important to note that, like composite functions, the order of transformations is B followed by A.

Remember, too, that $\det(MN) = \det(M) \times \det(N)$; in other words, if $\det(M) = 6$ and $\det(N) = 0.5$, then the area of a shape undergoing transformation MN will be multiplied by 3.

Reflections in lines which don't pass through the origin, and rotations about points other than the origin, can be generated by matrix transformations of the form:

$$\begin{pmatrix} x' \\ y' \end{pmatrix} = \begin{pmatrix} a & b \\ c & d \end{pmatrix}\begin{pmatrix} x \\ y \end{pmatrix} + \begin{pmatrix} e \\ f \end{pmatrix}.$$

For example, consider the transformation $\begin{pmatrix} 0 & 1 \\ 1 & 0 \end{pmatrix}\begin{pmatrix} x \\ y \end{pmatrix} + \begin{pmatrix} -2 \\ 2 \end{pmatrix}$. The unit square is transformed as follows: $\begin{pmatrix} 0 & 1 & 0 & 1 \\ 0 & 0 & 1 & 1 \end{pmatrix} \rightarrow \begin{pmatrix} -2 & -2 & -1 & -1 \\ 2 & 3 & 2 & 3 \end{pmatrix}$, and the diagram shows that the composite transformation reflection in $y = x$ followed by translation $\begin{pmatrix} -2 \\ 2 \end{pmatrix}$ is equivalent to reflection in the line $y = x + 2$.

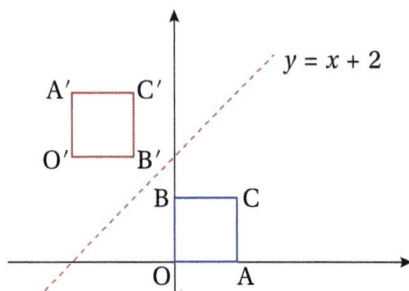

The transformation T is defined as $\begin{pmatrix} x' \\ y' \end{pmatrix} = \begin{pmatrix} 1 & 2 \\ 3 & -a \end{pmatrix}\begin{pmatrix} x \\ y \end{pmatrix} + \begin{pmatrix} b \\ 4a \end{pmatrix}$. P is the point $(2, a)$.

T transforms P to the point $(3, 9)$.

    (a)   Show that a possible solution is $a = 1$ and $b = -1$

    (b)   Find the other possible values of a and b.

    (c)   Write down the two possible points P.

(a) $\begin{pmatrix} 1 & 2 \\ 3 & -1 \end{pmatrix}\begin{pmatrix} 2 \\ 1 \end{pmatrix} + \begin{pmatrix} -1 \\ 4 \end{pmatrix} = \begin{pmatrix} 4 \\ 5 \end{pmatrix} + \begin{pmatrix} -1 \\ 4 \end{pmatrix} = \begin{pmatrix} 3 \\ 9 \end{pmatrix}$

(b) $\begin{pmatrix} 3 \\ 9 \end{pmatrix} = \begin{pmatrix} 1 & 2 \\ 3 & -a \end{pmatrix}\begin{pmatrix} 2 \\ a \end{pmatrix} + \begin{pmatrix} b \\ 4a \end{pmatrix}$

    (1) $3 = 2 + 2a + b$
    (2) $9 = 6 - a^2 + 4a$

    (2) $a^2 - 4a + 3 = 0 \Rightarrow a = 1$ or 3
    So the other value of $a = 3$.
    (1) $3 = 2 + 6 + b \Rightarrow b = -5$

(c)   So P = $(2, 1)$ or $(2, 3)$

(a) The command term "show that" means that substituting the values of a and b is a valid method of solution.

For part (b) we must create two equations by equating the top line and the bottom line of the matrix equation separately. Two unknowns need two equations.

Inverse matrices: If a matrix represents a transformation, then the inverse matrix represents the inverse transformation.

Inverse matrices are covered on page 21

For example, if $R = \begin{pmatrix} -1 & 0 \\ 0 & 1 \end{pmatrix}$ then $R^{-1} = \frac{1}{-1}\begin{pmatrix} 1 & 0 \\ 0 & -1 \end{pmatrix} = \begin{pmatrix} -1 & 0 \\ 0 & 1 \end{pmatrix}$. R is therefore self-inverse, and represents the transformation reflection in the y-axis, which is also self-inverse.

Example: The transformation with matrix $M = \begin{pmatrix} 2 & -1 \\ -3 & 2 \end{pmatrix}$ transforms the point P(x, y) to the point P$' = (3.5, -4.5)$. Find P.

Solution: Effectively, we have an equation of the form $MX = A$, and this can be solved by calculating $X = M^{-1}A$. So in this case, $\begin{pmatrix} x \\ y \end{pmatrix} = M^{-1}\begin{pmatrix} 3.5 \\ -4.5 \end{pmatrix}$ and using matrix functionality on the GDC we get P $= (2.5, 1.5)$

Matrix Transformations: Practice Exercise

Answers

1. A linear transformation T has matrix $\begin{pmatrix} 2 & -1 \\ 1 & 1 \end{pmatrix}$. Find the image of the point $(3, 1)$ under T, and the point having an image of $(7, 2)$.

2. The matrix $\begin{pmatrix} a & b \\ c & d \end{pmatrix}$ transforms $(1, 2)$ to $(3, 3)$ and $(-1, 1)$ to $(-3, 3)$. Set up two pairs of simultaneous equations and hence find the matrix.

3. Write down the matrices representing:

    $R =$ A reflection in the y-axis

    $S =$ A 90° clockwise rotation about the origin.

    $T =$ A reflection in the line $y = x$.

    Show by matrix multiplication that transformation R followed by transformation S is equivalent to transformation T. Determine whether S followed by R is also equivalent to T.

4. The matrix transformation $\begin{pmatrix} x' \\ y' \end{pmatrix} = \begin{pmatrix} 0 & -1 \\ 1 & 0 \end{pmatrix}\begin{pmatrix} x \\ y \end{pmatrix} + \begin{pmatrix} 8 \\ 2 \end{pmatrix}$ represents a +90° rotation about the point P $= (p, q)$. Find P.

    The centre of rotation is the only invariant point under the transformation.

Answers
1. $(5, 4); (3, -1)$
2. $\begin{pmatrix} 3 & 0 \\ -1 & 2 \end{pmatrix}$
3. $RS \neq SR$
4. P $= (3, 5)$

3.7 Basics of Vectors

Notation: Think of a vector as representing a movement, or displacement, in a plane. This can be represented by an arrow. The vector can be defined in several ways:

A vector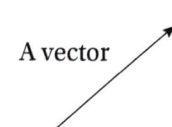

- Using a single small letter. Bold type in printed text, line underneath or arrow on top in handwriting.

- Using the named points at either end, arrow on top.

- Using a "column vector" to show the displacement in the x, y and z (if applicable) directions.

- In the form $a\boldsymbol{i} + b\boldsymbol{j}$ where $\boldsymbol{i}$ and $\boldsymbol{j}$ are unit vectors in the x and y directions (this is equivalent to the column vector form but less easy to use).

Same length, different direction – so a different vector

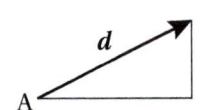

This vector could be written as:

$$\overrightarrow{AB}, \boldsymbol{d}, \binom{6}{3}, 6\boldsymbol{i} + 3\boldsymbol{j} \text{ (as examples)}$$

The column vector form is generally easier to use than the unit vector form. If a question uses unit vectors, I suggest you convert straight away to column vectors. The column vector form is also useful because we can work out the length and direction of the vector using Pythagoras and arctan (or using appropriate calculator functions).

Position and displacement vectors: If a vector is used to define the position of a point then it is known as a *position vector*. It will always start at the origin. The components of the column vector will always be the same as the coordinates of the point. *Displacement* vectors differ from position vectors in that they have no specific position – they just represent a **change** in position.

Operating with vectors: If you move along a vector $\boldsymbol{a}$ then along a vector $\boldsymbol{b}$, the single displacement which takes you to the end position is defined as vector $\boldsymbol{a} + \boldsymbol{b}$.

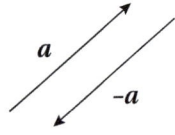

- The length of $\boldsymbol{a} + \boldsymbol{b}$ is not the length of $\boldsymbol{a}$ plus the length of $\boldsymbol{b}$; it is shorter. However, if $\boldsymbol{a}$ and $\boldsymbol{b}$ are written as column vectors, then adding them will give vector $\boldsymbol{a} + \boldsymbol{b}$. For example: $\binom{6}{-2} + \binom{-4}{5} = \binom{2}{3}$.

- When adding two or more vectors together, the single vector representing the sum is known as the *resultant*.

- A vector can be multiplied by a number. For example, $2\boldsymbol{a}$ has the same direction as $\boldsymbol{a}$ but is twice as long. Using column vectors, eg: $2\binom{3}{5} = \binom{6}{10}$.

- A minus sign reverses the direction of a vector.

Vector subtraction: To get from A to B using vectors, the path is $-\boldsymbol{a} + \boldsymbol{b}$ or $\boldsymbol{b} - \boldsymbol{a}$. Thus the vector $\overrightarrow{AB} = \boldsymbol{b} - \boldsymbol{a}$. This general principle should be remembered. It also works with position vectors.

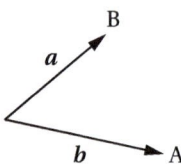

Alternatively:

$\overrightarrow{OA} = 4\boldsymbol{i} + 3\boldsymbol{j}, \overrightarrow{OB} = 6\boldsymbol{i} - \boldsymbol{j}$

$\overrightarrow{AB} = \overrightarrow{OB} - \overrightarrow{OA}$

$= (6\boldsymbol{i} - \boldsymbol{j}) - (4\boldsymbol{i} + 3\boldsymbol{j})$

$= (2\boldsymbol{i} - 4\boldsymbol{j})$

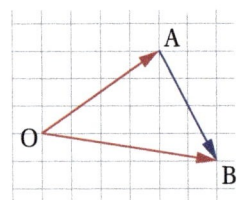

For example, if A is (4, 3) and B is (6, –1) then A and B have position vectors $\binom{4}{3}$ and $\binom{6}{-1}$.

So vector $\overrightarrow{AB} = \boldsymbol{b} - \boldsymbol{a} = \binom{6}{-1} - \binom{4}{3} = \binom{2}{-4}$.

A quadrilateral OABC has points O(0, 0) and A(4, 1).

(a) If $\overrightarrow{AB} = \begin{pmatrix} 1 \\ -2.5 \end{pmatrix}$, find the coordinates of B.

(b) (i) $\overrightarrow{OC} = \begin{pmatrix} -1 \\ k \end{pmatrix}$. Find k such that $\overrightarrow{CB}$ is parallel to $\overrightarrow{OA}$.

    (ii) What is the ratio of the lengths of OA to CB?

(a) $\overrightarrow{OB} = \overrightarrow{OA} + \overrightarrow{AB} = \begin{pmatrix} 4 \\ 1 \end{pmatrix} + \begin{pmatrix} 1 \\ -2.5 \end{pmatrix} = \begin{pmatrix} 5 \\ -1.5 \end{pmatrix}$

    B = (5, –1.5)

(b) (i) $\overrightarrow{CB} = \overrightarrow{OB} - \overrightarrow{OC} = \begin{pmatrix} 5 \\ -1.5 \end{pmatrix} - \begin{pmatrix} -1 \\ k \end{pmatrix}$

    $= \begin{pmatrix} 6 \\ -1.5 - k \end{pmatrix}$

    $\begin{pmatrix} 6 \\ -1.5 - k \end{pmatrix} = m\begin{pmatrix} 4 \\ 1 \end{pmatrix}$

    So $m = 1.5 \Rightarrow -1.5 - k = 1.5$

    $k = -3$

  (ii) $|\overrightarrow{OA}| = \sqrt{4^2 + 1^2} = \sqrt{17}$

    $\overrightarrow{CB} = \begin{pmatrix} 5 \\ -1.5 \end{pmatrix} - \begin{pmatrix} -1 \\ -3 \end{pmatrix} = \begin{pmatrix} 6 \\ 1.5 \end{pmatrix}$

    $|\overrightarrow{CB}| = \sqrt{6^2 + 1.5^2} = \sqrt{38.25}$

    $\therefore \overrightarrow{OA}:\overrightarrow{CB} = \sqrt{17} : \sqrt{38.25} = 1:1.5$

It's possible to become easily confused in questions like this. To avoid confusion, be rigorous with working. In particular, don't mix up the notation for vectors with the notation for coordinates. And always write down "what am I working out, how am I working it out, what is the answer?" In (b)(i), for example:

- *I'm working out vector CB*
- *I need to subtract OC from OB*
- *...and I get the answer.*

I've introduced letter m since parallel vectors are always multiples of each other. Once you know k you can work out the coordinates of all the points, and can then answer anything the examiner throws at you!

Three dimensional vectors: Conventional geometry is harder in three dimensions than in two: using vectors, the routines and calculations are pretty much the same in any number of dimensions – even 4! Thus vectors provide us with a powerful framework for solving problems in 3-d geometry.

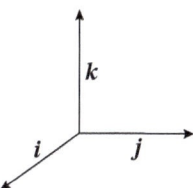

It is always hard to draw accurate diagrams of 3-d vectors. If it **has** to be done, the configuration as shown is often used, but others are possible.

Magnitude of vectors: The *magnitude* (or length) can be found by using Pythagoras' Theorem in two or three dimensions as appropriate.

The length of $2\boldsymbol{i} + 3\boldsymbol{j} - 5\boldsymbol{k}$ is $\sqrt{2^2 + 3^2 + (-5)^2} = \sqrt{38}$. Use the modulus sign to represent magnitude. Thus, $|\boldsymbol{p}|$ is the magnitude of $\boldsymbol{p}$.

Collinearity: If three points A, B and C are collinear (on the same straight line) then vectors $\overrightarrow{AB}$ and $\overrightarrow{BC}$ must be in the same direction and hence multiples (and $\overrightarrow{AC}$ will be multiples of both of them as well). This provides a simple test for collinearity. Consider the points A(2, 0, 3), B(3, 6, –1) and C(4.5, 15, –7):

$\overrightarrow{AB}$ is $\begin{pmatrix} 1 \\ 6 \\ -4 \end{pmatrix}$ and $\overrightarrow{BC}$ is $\begin{pmatrix} 1.5 \\ 9 \\ -6 \end{pmatrix}$. Thus $\overrightarrow{BC} = 1.5 \times \overrightarrow{AB}$, and therefore A, B and C are collinear.

Unit vectors: Any vector with length 1 is a *unit vector*. The three vectors $\boldsymbol{i}$, $\boldsymbol{j}$ and $\boldsymbol{k}$ are unit vectors which form the basis of the 3-d coordinate system. As with 2-d vectors, it is easier to use the column format eg: $\begin{pmatrix} 2 \\ 3 \\ 4 \end{pmatrix}$ rather than $2\boldsymbol{i} + 3\boldsymbol{j} + 4\boldsymbol{k}$. To find a unit vector in a particular direction, divide the given vector by its length. The symbol for a unit vector is $\hat{v}$

so $\hat{v} = \dfrac{v}{|v|}$.

Example: Find the unit vector in the direction $\overrightarrow{AB}$ where A is $(4, -1, -2)$ and B is $(6, 0, -4)$.

Solution: $\overrightarrow{AB} = \begin{pmatrix} 6 \\ 0 \\ -4 \end{pmatrix} - \begin{pmatrix} 4 \\ -1 \\ -2 \end{pmatrix} = \begin{pmatrix} 2 \\ 1 \\ -2 \end{pmatrix}$. Thus $|\overrightarrow{AB}| = \sqrt{2^2 + 1^2 + (-2)^2} = \sqrt{9} = 3$.

The unit vector is therefore $\dfrac{1}{3} \begin{pmatrix} 2 \\ 1 \\ -2 \end{pmatrix} = \begin{pmatrix} \frac{2}{3} \\ \frac{1}{3} \\ -\frac{2}{3} \end{pmatrix}$.

A vector can be rescaled using the unit vector. For example, find the vector with length 6 in the direction $4i + 3j$. $|4i + 3j|$ has length $\sqrt{4^2 + 3^2} = 5$, so the unit vector in the direction is $\frac{1}{5}(4i + 3j)$. Therefore the vector with length 6 in that direction is $\frac{6}{5}(4i + 3j) = \frac{24}{5}i + \frac{18}{5}j$.

3.8 Scalar (Dot) Product

Definition: The scalar product is a number which can be calculated from two vectors. On its own it has no real significance, but is used particularly in connection with angles between vectors. The scalar product of two vectors (2-d or 3-d) a and b is defined as: $a.b = |a||b|\cos\theta$ where θ is the angle between the directions of the two vectors. This formula can be read as: "The dot product of vectors a and b = the length of a times the length of b times the cosine of the angle between them." If the vectors are defined in column form, an alternative way of calculating the scalar product is:

$$\begin{pmatrix} a_1 \\ a_2 \\ a_3 \end{pmatrix} \cdot \begin{pmatrix} b_1 \\ b_2 \\ b_3 \end{pmatrix} = a_1 b_1 + a_2 b_2 + a_3 b_3.$$

Angle between two vectors: The dot product provides a convenient way of calculating the angle between two vectors. This is very common in exams, either as a short question or as part of a longer one. For example, to find the angle between $a = 2i + 3j - k$ and $b = 4i - 2j + 3k$:

$$|a| \text{ is } \sqrt{2^2 + 3^2 + (-1)^2} = \sqrt{14}; \text{ and } |b| \text{ is } \sqrt{4^2 + (-2)^2 + 3^2} = \sqrt{29}.$$

So $a.b = \sqrt{14}\sqrt{29} \cos\theta$. But $a.b$ can also be calculated using column vectors:

$$a.b = \begin{pmatrix} 2 \\ 3 \\ -1 \end{pmatrix} \cdot \begin{pmatrix} 4 \\ -2 \\ 3 \end{pmatrix} = 8 - 6 - 3 = -1$$

So, $\sqrt{14}\sqrt{29} \cos\theta = -1 \Rightarrow \cos\theta = -0.0496$, and finally we get the angle θ to be $\arccos(-0.0496) = 92.8°$. If the question asks for the acute angle between the corresponding *lines*, we must then give the answer $87.2°$.

Triangle RST has vertices at R(1, –1, 4), S(2, –1, 0), T(0, 1, 1)

    (a)   Find angle R.

    (b)   Hence find the area of the triangle.

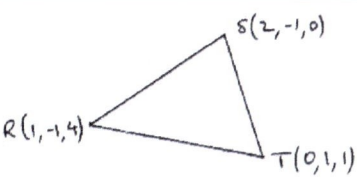

In 3-d vector questions, I often find it useful to draw a quick sketch. The sketch won't show points in actual positions, but it helps to envisage what's going on.

The vector directions are important. Angle R is between vectors $\vec{RS}$ and $\vec{RT}$ so using $\vec{SR}$, for example, will give the wrong answer (by 180°).

(a) $\vec{RS} = \begin{pmatrix} 1 \\ 0 \\ -4 \end{pmatrix}$; $\vec{RT} = \begin{pmatrix} -1 \\ 2 \\ -3 \end{pmatrix}$;

$\vec{RS}.\vec{RT} = -1 + 0 + 12 = 11$

$\sqrt{17}\,\sqrt{14}\cos R = 11$

$\cos R = \dfrac{11}{\sqrt{17}\,\sqrt{14}}$

$R = 44.5°$

(b)   Area $= \frac{1}{2}\,RS \times RT \times \sin 44.5 = 5.41$

An important property of the dot product is that, since $\cos 90° = 0$, it follows that if $a \perp b$ then $a.b = 0$ (and vice versa).

Example: If $p = 3i + 2aj - k$ and $q = i + (a - 3)j + (3a - 1)k$, find the possible values of a such that p is perpendicular to q.

Solution: $\begin{pmatrix} 3 \\ 2a \\ -1 \end{pmatrix}.\begin{pmatrix} 1 \\ a - 3 \\ 3a - 1 \end{pmatrix} = 0 \Rightarrow 3 + 2a^2 - 6a - 3a + 1 = 0$

$2a^2 - 9a + 4 = 0 \Rightarrow (2a - 1)(a - 4) = 0$

Thus $a = 0.5$ or 4.

Vector component:

In the diagram, the component of vector a in the direction of vector b is indicated by the red line – this could, for example, be the component of a force in a particular direction. The magnitude of the component (found by trigonometry) is $|a|\cos\theta$ and, since $a.b = |a||b|\cos\theta$, it follows that the magnitude of the component can be calculated as $\dfrac{a.b}{|b|}$.

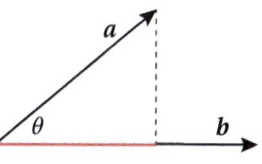

3.9 Vector (Cross) Product

Definition: Unlike the scalar product, the vector product of two vectors is itself a vector. The vector product is written as $a \times b$ and both a and b have to be vectors. If the vectors are written in column form, then the vector product is:

$$a \times b = \begin{pmatrix} a_1 \\ a_2 \\ a_3 \end{pmatrix} \times \begin{pmatrix} b_1 \\ b_2 \\ b_3 \end{pmatrix} = \begin{pmatrix} a_2 b_3 - a_3 b_2 \\ a_3 b_1 - a_1 b_3 \\ a_1 b_2 - a_2 b_1 \end{pmatrix}$$

Look carefully at the patterns to see how to memorise them.

Applications: The main property of the vector $\boldsymbol{a} \times \boldsymbol{b}$ is that it is perpendicular to both $\boldsymbol{a}$ and $\boldsymbol{b}$.

Use the scalar product to test the answer.

For example, if $\boldsymbol{a} \times \boldsymbol{b} = \begin{pmatrix} 2 \\ 1 \\ -3 \end{pmatrix} \times \begin{pmatrix} 0 \\ -2 \\ 4 \end{pmatrix} = \begin{pmatrix} 4-6 \\ 0-8 \\ (-4)-0 \end{pmatrix} = \begin{pmatrix} -2 \\ -8 \\ -4 \end{pmatrix}$, then $\begin{pmatrix} -2 \\ -8 \\ -4 \end{pmatrix}$ is perpendicular

to $\boldsymbol{a}$ and $\boldsymbol{b}$.

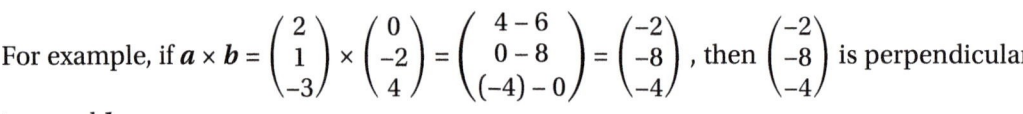

Another useful property is that $|\boldsymbol{a} \times \boldsymbol{b}| = |\boldsymbol{a}||\boldsymbol{b}|\sin\theta$ where θ is the angle between the two vectors. This expression is equivalent to the area of the parallelogram which can be formed from the two vectors, and also leads to the formula for the area of a triangle formed from the two vectors: Area $= \frac{1}{2}|\boldsymbol{a} \times \boldsymbol{b}|$. This gives us an alternative method for finding the area of the triangle in the exam-style question in the previous section:

$$\overrightarrow{RS} \times \overrightarrow{RT} = \begin{pmatrix} 1 \\ 0 \\ -4 \end{pmatrix} \times \begin{pmatrix} -1 \\ 2 \\ -3 \end{pmatrix} = \begin{pmatrix} 0-(-8) \\ 4-(-3) \\ 2-0 \end{pmatrix} = \begin{pmatrix} 8 \\ 7 \\ 2 \end{pmatrix}$$

$$\left|\overrightarrow{RS} \times \overrightarrow{RT}\right| = \sqrt{8^2 + 7^2 + 2^2} = \sqrt{117}$$

$$\text{Area of triangle } = \frac{1}{2}\sqrt{117} = 5.41$$

Vector component: Similar to the component of a vector in the direction of another vector (see previous section) the cross product can be used to find the component of a vector in a direction **perpendicular** to another vector. This can be calculated as $\dfrac{|\boldsymbol{a} \times \boldsymbol{b}|}{|\boldsymbol{b}|}$.

3.10 Equations of Lines

Vector equation of a line: Although we can use the Cartesian equation of a line in two dimensions ($y = mx + c$), in three dimensions it becomes very unwieldy (see below). The *vector equation* of a line is much easier to use, and has the same form in both two and three dimensions.

The line shown in the diagram has a direction given by the vector $\begin{pmatrix} 2 \\ -1 \end{pmatrix}$. We can find the position vector of any point on the line by first going along a vector which **takes** us to the line – say, to the point (2, 4) – and then adding any multiple of the direction vector. This gives us the vector equation of the line.

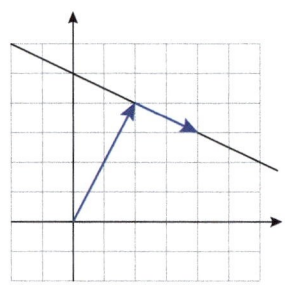

In this case it would be: $\boldsymbol{r} = \begin{pmatrix} 2 \\ 4 \end{pmatrix} + t\begin{pmatrix} 2 \\ -1 \end{pmatrix}$.

- The r indicates the position vector of a general point on the line and could also be written as $\begin{pmatrix} x \\ y \end{pmatrix}$.

- The vector $\begin{pmatrix} 2 \\ 4 \end{pmatrix}$ is the position of a point on the line – any other point on the line could have been used.

Any letter can be used for the parameter, including Greek letters. λ is often used.

- t is called the *parameter*. Different values of t give us different points. For example, if $t = 2$, we get the point (6, 2), and every point on the line corresponds to a particular value of t.

- The vector $\begin{pmatrix} 2 \\ -1 \end{pmatrix}$ is the *direction vector* of the line. Other multiples, such as $\begin{pmatrix} 4 \\ -2 \end{pmatrix}$ or $\begin{pmatrix} -2 \\ 1 \end{pmatrix}$ could have been used.

In general, the vector equation of a line is $r = a + \lambda b$ where a is the position vector of a point on the line and b is a direction vector. The only difference in 3 dimensions is that r, a and b are 3-d vectors.

Example: Find a vector equation of the line passing through $(-3, 4, 1)$ and $(2, -2, 2)$.

Solution: The direction vector of the line is $\begin{pmatrix} 2 \\ -2 \\ 2 \end{pmatrix} - \begin{pmatrix} -3 \\ 4 \\ 1 \end{pmatrix} = \begin{pmatrix} 5 \\ -6 \\ 1 \end{pmatrix}$. We could now use either point as the position vector – I tend to use the one with the simpler numbers if there's a choice.

Thus a vector equation of the line is $r = \begin{pmatrix} 2 \\ -2 \\ 2 \end{pmatrix} + \lambda \begin{pmatrix} 5 \\ -6 \\ 1 \end{pmatrix}$.

> It also follows that a general point on the line could be written as: $(2 + 5\lambda, -2 - 6\lambda, 2 + \lambda)$.

3.11 Application to Kinematics

Consider the position of an object given by the vector equation $r = \begin{pmatrix} 2 \\ 0 \end{pmatrix} + t \begin{pmatrix} 0.7 \\ 1 \end{pmatrix}$, where t is the time in seconds since the start of the motion. Clearly the initial position is the point $(2, 0)$ and, since the object will change position by $\begin{pmatrix} 0.7 \\ 1 \end{pmatrix}$m every second, this vector represents the velocity. In other words, the object is travelling in a straight line with constant velocity. Its speed is the magnitude of the velocity vector, which is $\sqrt{0.7^2 + 1^2} = 1.22\,\text{ms}^{-1}$.

> In general, $r = r_0 + tv$.

Intersection of paths: If two objects are moving in straight lines at constant speeds, we can find the point of intersection of their paths (if there is one). Consider two model planes, P_1 and P_2. Their paths relative to a fixed origin on the ground are given by the vector equations: $r_1 = \begin{pmatrix} 1 \\ -5 \\ 0 \end{pmatrix} + t \begin{pmatrix} -1 \\ 3 \\ 1 \end{pmatrix}$ and $r_2 = \begin{pmatrix} -11 \\ 1 \\ 0 \end{pmatrix} + s \begin{pmatrix} 0 \\ 5 \\ 2 \end{pmatrix}$. We can find the point of intersection of the paths by equating coordinates:

> Note that the z coordinate measures height and, in both cases, is 0 at time 0.

x: $\quad 1 - t = -11$

y: $\quad -5 + 3t = 1 + 5s$

z: $\quad t = 2s$

We only need two of the equations to find the solution – let's use those for x and z. These give $t = 12$ and $s = 6$. Check that these work in the equation for y (if they didn't, then that implies there is no point of intersection). Now substituting either of s or t into the vector equations gives the point of intersection as $(-11, 31, 12)$.

The question arises: do the planes collide? Answer: only if they are both at the point of intersection at the same time. If they took off simultaneously then they won't collide since P_1 arrives after 12 s and P_2 after 6 s. But if P_1 took off 6 s after P_2 then they would collide.

In this question, a unit vector represents a displacement of 1 metre. A miniature car moves in a straight line, starting at the point $(2, 0)$. After t seconds, its position (x, y) is given by the vector equation $\begin{pmatrix} x \\ y \end{pmatrix} = \begin{pmatrix} 0 \\ 2 \end{pmatrix} + t\begin{pmatrix} 2 \\ 1 \end{pmatrix}$.

(a) How far is the car from the point $(0, 0)$ after 2 seconds.

(b) Find the speed of the car.

(c) Obtain the equation of the car's path in the form $ax + by = c$.

Another miniature vehicle, a motorcycle, starts at the point $(6, -2)$ four seconds after the car and travels in a straight line with constant speed. The equation of its path is $5x - 3y = d$.

(d) (i) Find the value of d

(ii) Find the coordinates of the point where the two paths cross.

(iii) Find the vector equation for the position of the motorcycle, given that the car and the motorcycle collide.

(a) At $t = 2$, $\begin{pmatrix} x \\ y \end{pmatrix} = \begin{pmatrix} 0 \\ 2 \end{pmatrix} + 2\begin{pmatrix} 2 \\ 1 \end{pmatrix} = \begin{pmatrix} 4 \\ 4 \end{pmatrix}$

From $(0, 0)$ to $(4, 4)$ the distance is $\sqrt{32}$

(b) The speed of the car is $\sqrt{2^2 + 1^2} = \sqrt{5}\,\text{ms}^{-1}$

(c) $x = 2t$

$y = 2 + t \Rightarrow 2t = 2y - 4$

$\therefore\ x = 2y - 4 \Rightarrow x - 2y = -4$

(d) (i) At $(6, -2)$, $5 \times 6 - 3 \times (-2) = d \Rightarrow d = 36$

(ii) Solve $x - 2y = -4$ and $5x - 3y = 36$ simultaneously

$x = 12, y = 8$ (GDC) $\therefore$ Paths cross at $(12, 8)$

(iii) Car reaches $(12, 8)$ at $t = 6$

Motorcycle vector equation is:

$$\begin{pmatrix} x \\ y \end{pmatrix} = \begin{pmatrix} 6 \\ -2 \end{pmatrix} + (t - 4)\begin{pmatrix} v_x \\ v_y \end{pmatrix}$$

At $t = 6$, $\begin{pmatrix} 12 \\ 8 \end{pmatrix} = \begin{pmatrix} 6 \\ -2 \end{pmatrix} + 2\begin{pmatrix} v_x \\ v_y \end{pmatrix} \Rightarrow \begin{pmatrix} v_x \\ v_y \end{pmatrix} = \begin{pmatrix} 3 \\ 5 \end{pmatrix}$

$\therefore$ Vector equation of motorcycle path is:

$$\begin{pmatrix} x \\ y \end{pmatrix} = \begin{pmatrix} 6 \\ -2 \end{pmatrix} + (t - 4)\begin{pmatrix} 3 \\ 5 \end{pmatrix} = \begin{pmatrix} -6 \\ -22 \end{pmatrix} + t\begin{pmatrix} 3 \\ 5 \end{pmatrix}$$

Parts (a), (b) and (c) are relatively straightforward, although note how I have eliminated in t in part (c). It could have been done in other ways, but I like to keep everything in whole numbers where possible.

The key point in (c)(iii) is that the car arrives at the collision point at t = 6 and, since the motorcycle sets off 4 seconds later, it must get there at s = 2.

Closest point calculations: In two dimensions the non-parallel paths of two objects will always cross. This is not necessarily the case in three dimensions, but we can still calculate the closest distance between them.

So let us suppose that the motion of two objects is given by the following vector equations:

$$r_1 = \begin{pmatrix} 1 \\ 1 \\ -2 \end{pmatrix} + t\begin{pmatrix} 0 \\ 1 \\ 1 \end{pmatrix}, \quad r_2 = \begin{pmatrix} 0 \\ 7 \\ -1 \end{pmatrix} + s\begin{pmatrix} 2 \\ -1 \\ 0 \end{pmatrix}$$

We shall consider the situation where the two objects set off at the same time – when are they closest to each other?

We can use t as the parameter in both vector equations, in which case the general vector between their two positions becomes $\begin{pmatrix} 1 \\ 1 + t \\ -2 + t \end{pmatrix} - \begin{pmatrix} 2t \\ 7 - t \\ -1 \end{pmatrix} = \begin{pmatrix} 1 - 2t \\ -6 \\ -1 + t \end{pmatrix}$.

In other words, at any time t we can find the vector which gives the relative position of the two objects.

The length of this vector is $\sqrt{(1 - 2t)^2 + (-6)^2 + (-1 + t)^2}$ and we can find the minimum distance by differentiation. By entering the function on the GDC, you should find that the minimum occurs at $t = 0.6$ at which time the separation between the objects is 6.02. You might like to try the differentiation by hand as well – note that to find the value of t it is only necessary to differentiate the function inside the square root.

Full working on the website.

Variable velocity: In the Calculus chapter, the section on kinematics shows that if displacement is given as a function of time, $f(t)$, then velocity is given by $f'(t)$ and acceleration by $f''(t)$. This can also be applied to displacements given as vectors (although you will not be given such problems in three dimensions). For example, if an object has

See page 191

displacement $r = \begin{pmatrix} 3t^2 - 2 \\ 4 - 6t \end{pmatrix}$, then $v = \begin{pmatrix} 6t \\ -6 \end{pmatrix}$ and $a = \begin{pmatrix} 6 \\ 0 \end{pmatrix}$, thus showing that the acceleration is constant and parallel to the x-axis.

Example: A particle's velocity t seconds after measurements begin is given by $v = \begin{pmatrix} 2t - 2 \\ 3 \end{pmatrix}$. Given that its initial position is $(3, 1)$, find an expression for its displacement at time t. Hence find its distance from $(0, 0)$ and its speed at $t = 3$.

Solution: By integration, $r = \begin{pmatrix} t^2 - 2t \\ 3t \end{pmatrix} + \begin{pmatrix} c_1 \\ c_2 \end{pmatrix}$. When $t = 0$, $r = \begin{pmatrix} 3 \\ 1 \end{pmatrix}$ thus the displacement is $r = \begin{pmatrix} t^2 - 2t + 3 \\ 3t + 1 \end{pmatrix}$. When $t = 3$, $r = \begin{pmatrix} 6 \\ 10 \end{pmatrix}$ and $v = \begin{pmatrix} 4 \\ 3 \end{pmatrix}$.

Distance from $(0, 0) = \sqrt{6^2 + 10^2} = \sqrt{136}$, and speed $= \sqrt{4^2 + 3^2} = 5$.

This shows how to deal with the constant of integration in a vector context.

Projectiles: In projectile motion it is assumed that an object's only acceleration is due to gravity, and is therefore constant. This means that the horizontal component of its velocity is constant. Let's take a case where a football is kicked with an initial velocity of $24\,\mathrm{ms}^{-1}$ at $30°$ to the horizontal. As a vector, this is $\begin{pmatrix} 24\cos30° \\ 24\sin30° \end{pmatrix} = \begin{pmatrix} 12\sqrt{3} \\ 12 \end{pmatrix}\,\mathrm{ms}^{-1}$

- Assuming the acceleration due to gravity is $10\,\mathrm{ms}^{-2}$, $a = \begin{pmatrix} 0 \\ -10 \end{pmatrix}$

- Then velocity at time t will be $v = \begin{pmatrix} 0 \\ -10t \end{pmatrix} + \begin{pmatrix} 12\sqrt{3} \\ 12 \end{pmatrix} = \begin{pmatrix} 12\sqrt{3} \\ -10t + 12 \end{pmatrix}$

- And the displacement will be $s = \begin{pmatrix} 12\sqrt{3}\,t \\ -5t^2 + 12t \end{pmatrix} + \begin{pmatrix} x_0 \\ y_0 \end{pmatrix}$ where $\begin{pmatrix} x_0 \\ y_0 \end{pmatrix}$ is the ball's initial position.

I've integrated acceleration to get velocity, then again to get displacement, not forgetting the constant (vector) of integration in each case.

Armed with these vectors we can find out anything we need to about the motion, usually by dealing with horizontal and vertical components separately. Let's take the initial position as the origin, so $s = \begin{pmatrix} 12\sqrt{3}\,t \\ -5t^2 + 12t \end{pmatrix}$

Q. ***When does the ball reach maximum height?***

A. When $v_y = 0$, so $-10t + 12 = 0 \Rightarrow t = 1.2\,\mathrm{s}$

I'm using s_x, s_y, v_x, v_y as the horizontal and vertical components of displacement and velocity.

Q. **What is the maximum height?**

A. When $t = 1.2$, $s_y = -5 \times 1.2^2 + 12 \times 1.2 = 7.2$ m

Q. **When does the ball hit the ground?**

A. When $s_y = 0$, so $-5t^2 + 12t = 0 \Rightarrow t = 2.4$ s

Q. **How far does it travel?**

A. When $t = 2.4$, $s_x = 12\sqrt{3} \times 2.4 = 49.9$ m

Q. **Will the ball clear a bar with height 4m placed 40m away?**

A. $s_x = 40 \Rightarrow t = 1.925$ s. When $t = 1.925$, $s_y = 4.57$, so the ball will clear the bar.

As ever, questions such as these can be twisted and turned – but just keep an eye on the components, and you should be able to see how to get to the answer. For example, the last question above could be changed to: "How far away should a 4 m bar be placed such that the ball just clears it on its way down?" The answer is 41.6 m.

Circular motion: With circular motion, don't confuse actual velocity (v) with angular velocity (ω). The angular velocity is the rate of change of angle; thus the angle turned from $t = 0$ is given by the angular velocity multiplied by the time.

- $\theta = \omega t$

The direction of the velocity is always at a tangent to the circle, and its magnitude is given by:

- $v = r\omega$

The direction of the acceleration is always towards the centre of the circle, and its magnitude is given by:

- $a = r\omega^2$

Note that ω is constant, and therefore so is the speed. But the velocity v is variable since its direction changes.

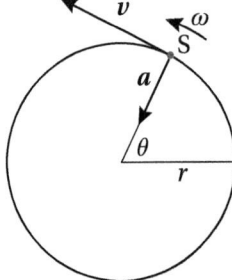

Note that $\boldsymbol{s}.\boldsymbol{v} = 0$. This is one method of showing that the velocity is tangential to the circle

It can be seen from the diagram that if an object S has displacement $\boldsymbol{s}$ at time t:

$$\boldsymbol{s} = \begin{pmatrix} r\cos\theta \\ r\sin\theta \end{pmatrix} = \begin{pmatrix} r\cos\omega t \\ r\sin\omega t \end{pmatrix}$$

$$\boldsymbol{v} = \begin{pmatrix} -v\sin\theta \\ v\cos\theta \end{pmatrix} = \begin{pmatrix} -v\sin\omega t \\ v\cos\omega t \end{pmatrix}$$

$$\boldsymbol{a} = \begin{pmatrix} -a\cos\theta \\ -a\sin\theta \end{pmatrix} = \begin{pmatrix} -r\omega^2\cos\omega t \\ -r\omega^2\sin\omega t \end{pmatrix}$$

The position vector of an object in circular motion is given by $\begin{pmatrix} x \\ y \end{pmatrix} = \begin{pmatrix} r\cos\omega t \\ r\sin\omega t \end{pmatrix}$ where r is the radius of the circle, ω is a constant and t is time.

(a) (i) Show that the velocity vector at time t is $v = \begin{pmatrix} -\omega r\sin\omega t \\ \omega r\cos\omega t \end{pmatrix}$.

(ii) Hence show that the speed is given by $|v| = \omega r$

The diagram shows the position of the city of Ankara on the Earth's surface, assumed to be spherical, at a latitude of 40°N. The equator has a circumference of 40 000 km.

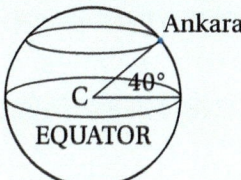

(b) (i) Find the radius of the Earth at the equator.

(ii) Hence find the radius of the Earth at latitude 40°.

Ankara revolves around a circle once every 24 hours.

(c) Show that the angular velocity of Ankara is 7.272×10^{-5} rad/s

(d) (i) Taking the centre of the circle as the origin, write down the displacement of Ankara as a vector in the form $\begin{pmatrix} r\cos\omega t \\ r\sin\omega t \end{pmatrix}$ where t is in seconds.

(ii) Find, by differentiation, the velocity vector of Ankara.

(iii) Hence find the speed of Ankara as it travels around the circle, giving your answer in ms⁻¹.

(b) (i) 6366 km

(ii) 4877 km

(d) (i) $\begin{pmatrix} 4877\cos 7.272 \times 10^{-5}\,t \\ 4877\sin 7.272 \times 10^{-5}\,t \end{pmatrix}$

(ii) $\begin{pmatrix} -0.3547\sin 7.272 \times 10^{-5}\,t \\ 0.3547\cos 7.272 \times 10^{-5}\,t \end{pmatrix}$

(iii) 354 ms⁻¹

Vectors: Practice Exercise

For questions 1–5, A = (1, 0, 2), B = (1, 1, 4), C = (0, 2, –2)

1. Find vector equations of lines (AB) and (AC).

2. Find the lengths of vectors $\overrightarrow{AB}$ and $\overrightarrow{AC}$.

3. Find the acute angle between lines (AB) and (AC).

4. Hence find the area of triangle ABC.

5. Now use the vector product to find the area of triangle ABC.

Answers

1. (AB): $r = \begin{pmatrix} 1 \\ 0 \\ 2 \end{pmatrix} + \lambda \begin{pmatrix} 0 \\ 1 \\ 2 \end{pmatrix}$

(AC): $r = \begin{pmatrix} 1 \\ 0 \\ 2 \end{pmatrix} + \lambda \begin{pmatrix} -1 \\ 2 \\ -4 \end{pmatrix}$

Other versions are possible.

2. $\sqrt{5}$, $\sqrt{21}$

3. 54.2°

4. 4.153

5. $\overrightarrow{AB} \times \overrightarrow{AC} = \begin{pmatrix} 8 \\ 2 \\ -1 \end{pmatrix}$

$\frac{1}{2}|\overrightarrow{AB} \times \overrightarrow{AC}| = 4.153$

6. (4, 6, 1)

8. $a = 3$

6. Line L_1 is given by $r = \begin{pmatrix} 0 \\ -2 \\ 5 \end{pmatrix} + t\begin{pmatrix} 1 \\ 2 \\ -1 \end{pmatrix}$

Line L_2 is given by $r = \begin{pmatrix} 1 \\ -3 \\ 4 \end{pmatrix} + t\begin{pmatrix} -1 \\ -3 \\ 1 \end{pmatrix}$ Find the point of intersection of L_1 and L_2.

7. Two objects are following straight line paths given by vector equations $r_1 = \begin{pmatrix} -1 \\ 4 \\ 3 \end{pmatrix} + s\begin{pmatrix} 1 \\ 0 \\ 3 \end{pmatrix}$ and $r_2 = \begin{pmatrix} 0 \\ 1 \\ 2 \end{pmatrix} + t\begin{pmatrix} 1 \\ 1 \\ 1 \end{pmatrix}$ with distances measured in metres. If the two objects start moving at the same time, show that the closest they come to each other is 3.29 m.

8. The position of a particle at a time t moving under a variable magnetic field is given by $r = \begin{pmatrix} at - 4 \\ t^2 + 2 \end{pmatrix}$ where $a > 0$. Given that the particle is moving at a speed of $\sqrt{13}$ when $t = 1$, find the value of a.

3.12 Graph Theory Basics

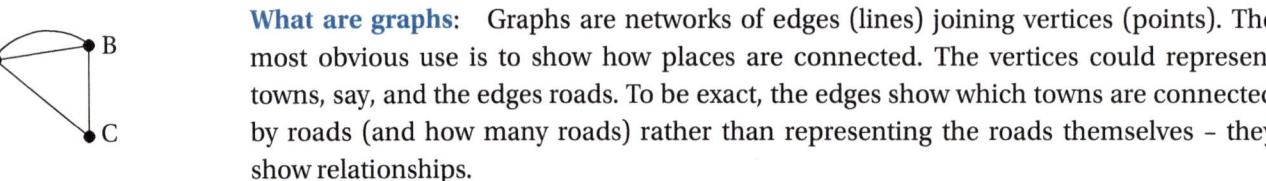

What are graphs: Graphs are networks of edges (lines) joining vertices (points). The most obvious use is to show how places are connected. The vertices could represent towns, say, and the edges roads. To be exact, the edges show which towns are connected by roads (and how many roads) rather than representing the roads themselves – they show relationships.

Sometimes some of the edges are *directed* – they are one-way. A graph containing directed edges is called a *directed graph* or *digraph*. The digraph illustrated could show:

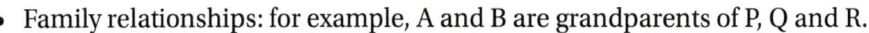

- Family relationships: for example, A and B are grandparents of P, Q and R.
- Skills: A and B are people, and the arrows indicate which of skills P, Q and R they possess.
- Transportation: A and B are depots, P, Q and R are retail outlets supplied by the depots.

Definitions: There is a considerable amount of graph terminology and you should understand exactly what each term means, and be able to use the terms in your answers.

Loop:	An edge that starts and ends at the same vertex.
Multiple edge:	When two vertices are joined by more than one edge.
Adjacent vertices:	Two vertices joined by an edge.
Adjacent edges:	Two edges with a common vertex.
Degree of a vertex:	Number of edges which meet at the vertex.
Indegree:	In a directed graph, the number of edges directed *in* to a vertex.
Outdegree:	In a directed graph, the number of edges directed *out* of a vertex.

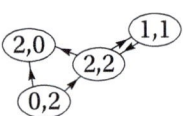

The diagram shows a directed network with vertices labelled (indegree, outdegree).

Types of graph:

Simple:	No loops, no multiple edges.
Connected:	Every vertex is accessible (not necessarily in one stage) from every other vertex.

Strongly connected: A directed graph is strongly connected if every vertex is accessible along a directed path from every other vertex. The directed network illustrated is connected – but not strongly connected because there are no paths from A to C or B to C.

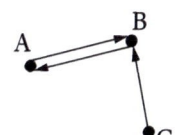

Complete: Every pair of vertices is adjacent (ie one move can get you from one vertex to any other vertex). A complete graph with *n* vertices is given the symbol K_n.

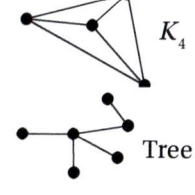

Tree: A simple connected graph which has no *cycles*; in other words, there are no round trips in a tree.

Weighted: In a weighted graph, the edges have numbers associated with them. The numbers could represent distances, travelling times, costs.

Subgraph: A graph whose vertices and edges are all contained in a larger graph. Note that a graph is a subgraph of itself.

3.13 Walks and Paths

Definitions: Once we start moving along the edges around a graph, we need to learn a whole new set of definitions:

Walk: A sequence of edges with no breaks.

Trail: A walk with no repeated edges.

Path: A trail with no repeated vertices.

Cycle: A path which ends where it starts.

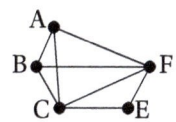

ABCFBC is a walk.
ABFCEFA is a trail.
ABFCE is a path.
ABCEFA is a cycle.

Hamiltonian and Eulerian graphs: Problems often involve travellers who want to create a tour of a set of towns, visiting each one just once; or inspectors who want to drive once along every road in a district.

Eulerian trail: An Eulerian trail on a connected graph is one which visits every edge of the graph exactly once. If the Eulerian trail starts and ends on the same vertex, it is called an *Eulerian circuit*.

It is easy to remember which is which: *E*ulerian graphs are to do with *E*dges.

Hamiltonian path: A Hamiltonian path on a connected graph visits every vertex exactly once. If the Hamiltonian path starts and ends on the same vertex, it is called a *Hamiltonian cycle*.

Thus an *Eulerian trail* can go through the same vertex more than once, even though no edges are repeated. It is the formal definition of those puzzles where you are asked if you can copy a shape without taking your pencil off the paper and without repeating any edge. In a *Hamiltonian path* there will be no repeated vertices *or* edges, but it is not necessary to include *every* edge.

Clearly if there were any repeated edges there would have to be repeated vertices as well.

The following graphs contain examples of Eulerian trails and circuits, and Hamiltonian paths and cycles. See if you can find relevant examples of each before looking at the answers below.

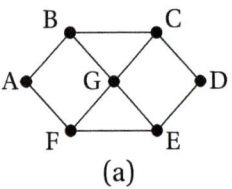
(a)

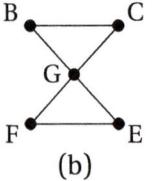
(b)

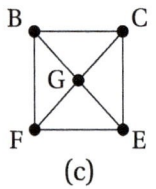
(c)

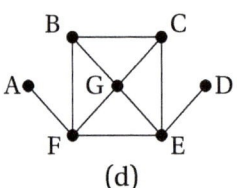
(d)

(a) Eulerian trail: ABCDEFBGECGFA (also a circuit)
     Hamiltonian path: ABCDEGFA (also a cycle)
(b) Eulerian trail: BCGFEGB (also a circuit)
     Hamiltonian path: FEGBC (but no cycles are possible)
(c) Eulerian trail: No. Always left with one edge unvisited.
     Hamiltonian path: BCEFGB (also a cycle)
(d) Eulerian trail: Not possible
     Hamiltonian path: AFBGCED (no cycles possible)

A graph with an Eulerian circuit is called *traversable*. A graph with an Eulerian trail is called *semi-traversable*.

The necessary and sufficient condition for there to be an Eulerian circuit is that all the vertices will have an even degree. To have just an Eulerian trail there will be exactly two vertices with odd degree. The condition for a graph to be Hamiltonian is beyond the scope of the course; any exam questions will be answered by inspection.

Graph	Even vertices	Odd vertices
a	7	0
b	5	0
c	1	4
d	1	6

For example: FBCBGECGFE

If we added another edge BC to graph (c), vertices B and C will become even, leaving just two odd vertices. There should now be an Eulerian trail – see if you can find one.

Note that if there are two odd vertices then an Eulerian trail will begin at one of them and end at the other.

The Königsberg bridges: The diagram shows a map of 18th-century Königsberg, with its seven bridges highlighted in yellow. Graph Theory began when in 1736 Euler was asked whether it was possible to walk between the various areas of the city, crossing each bridge once and once only.

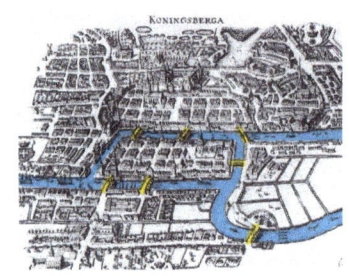

Solution: A graph can be drawn where each bridge is represented by an edge, and the vertices are the various land areas connected by the bridges:

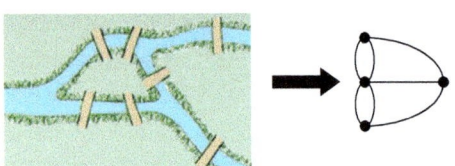

As can be seen, all four nodes have an odd degree, so no Eulerian trail is possible.

If a new bridge were to be built across the river to the left of the island, would that make a difference? Yes: the top and bottom nodes would become even, leaving two odd nodes. Thus a trail can be found, but not a circuit.

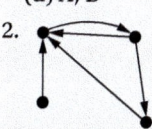

Graphs: Quick Practice

1. For the following graphs:

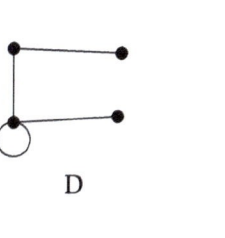

| A | B | C | D |

(a) Which are (i) Simple, (ii) Tree, (iii) Complete?

(b) List which graphs are subgraphs of other graphs.

(c) Which graphs contain loops?

(d) Which graphs contain multiple edges?

2. Draw a graph which has the following indegree-outdegree pairs:

$$((3, 1), (1, 2), (1, 1), (0, 1))$$

3. (a) Write down a Hamiltonian path for the graph on the right.

(b) Explain why there cannot be an Eulerian trail.

(c) Suggest an edge which can be added to the graph such that a Hamiltonian circuit exists.

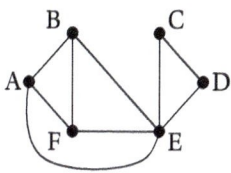

(d) Suggest an edge which can be added to the graph such that an Eulerian trail exists.

Answers

1. (a)(i) C, (ii) C, (iii) A, B

   (b) B ⊂ A, C ⊂ A, C ⊂ D

   (c) D

   (d) A, B

2.

3. (a) ABFEDC

   (b) 4 odd vertices

   (c) eg: BD (AFECDBA)

   (d) AB (reduces graph to two odd vertices)

3.14 Adjacency and Transition Matrices

Adjacency matrices: Since a graph shows which vertices are related to each other, the information can be shown equally well in an *adjacency matrix*. The adjacency matrix below (assuming rows and columns are labelled alphabetically) contains a 1 to indicate an edge connecting two vertices.

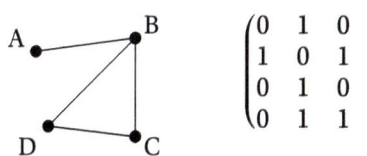

$$\begin{pmatrix} 0 & 1 & 0 & 0 \\ 1 & 0 & 1 & 1 \\ 0 & 1 & 0 & 1 \\ 0 & 1 & 1 & 0 \end{pmatrix}$$

If vertices i and j are connected, then so are j and i; as a result, the matrix is symmetrical. Note also that the leading diagonal is composed of zeroes – this will always be true for a simple graph. If any of the edges are directed the matrix will lose its symmetry.

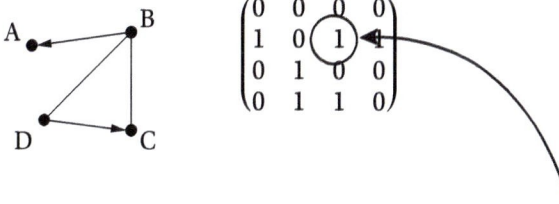

$$\begin{pmatrix} 0 & 0 & 0 & 0 \\ 1 & 0 & 1 & 1 \\ 0 & 1 & 0 & 0 \\ 0 & 1 & 1 & 0 \end{pmatrix}$$

The rows represent "from" and the columns "to". Thus, the circled element shows that an edge connects B to C.

The adjacency matrix shows the number of "1-stage" routes between vertices. Squaring the matrix will show us the number of "2-stage" routes.

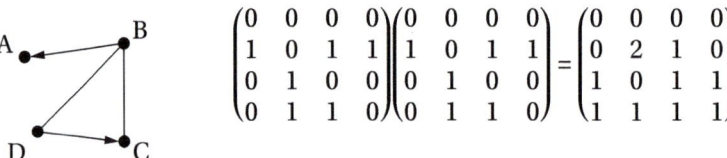

In general, for an adjacency matrix M, the matrix M^n will show the number of n-stage routes between vertices.

For example, although there is no 1-stage route from C to D, there is a 2-stage route (CBD); and there are two 2-stage routes from B to B (BCB, BDB).

Weighted adjacency matrices: When the edges of a graph are weighted (ie have numbers associated with them), then the adjacency matrix can show these weights rather than the number of edges. It is then known as a *weighted adjacency matrix*. Weighted adjacency matrices are only constructed for complete graphs: otherwise, if a pair of vertices were **not** connected by an edge, what entry would you put in the matrix? A zero would imply zero cost.

Example: Draw the graph defined by the weighted adjacency matrix A where each weight represents the cost of travelling between a network of towns. By trial and error, find the minimum cost of visiting each town, starting and ending at A.

$$\begin{array}{c} \\ A \\ B \\ C \\ D \end{array} \begin{array}{c} \begin{array}{cccc} A & B & C & D \end{array} \\ \begin{pmatrix} 0 & 3 & 2 & 5 \\ 3 & 0 & 5 & 4 \\ 2 & 5 & 0 & 6 \\ 5 & 4 & 6 & 0 \end{pmatrix} \end{array}$$

First construct a K_4 graph, then add the weights as shown by the matrix.

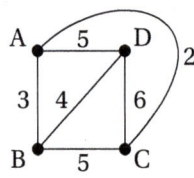

The lowest cost route from A back to A is ACDBA, a cost of 15 units.

Transition matrices: A *transition matrix* shows the probabilities of movement between one vertex and another. Consider the graph on the right.

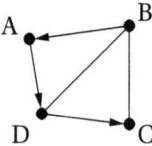

From vertex C the only possible movement is to B, giving rise to a probability of 1. However, from D, two movements are possible – to B and to C – and we therefore assign a probability of $\frac{1}{2}$ to each. The full transition matrix (where columns are "from" and rows are "to") is:

Note that this is the opposite way to an adjacency matrix.

$$T = \begin{pmatrix} 0 & \frac{1}{3} & 0 & 0 \\ 0 & 0 & 1 & \frac{1}{2} \\ 0 & \frac{1}{3} & 0 & \frac{1}{2} \\ 1 & \frac{1}{3} & 0 & 0 \end{pmatrix}. \text{ Note that the columns must add to give 1.}$$

Now if we were to make n moves, the matrix T^n will show us the long-term probability of being at each vertex.

In practice we cannot calculate this on the GDC, so we just use a large number for n, such as 50.

$$T^{50} = \begin{pmatrix} \frac{1}{8} & \frac{1}{8} & \frac{1}{8} & \frac{1}{8} \\ \frac{3}{8} & \frac{3}{8} & \frac{3}{8} & \frac{3}{8} \\ \frac{1}{4} & \frac{1}{4} & \frac{1}{4} & \frac{1}{4} \\ \frac{1}{4} & \frac{1}{4} & \frac{1}{4} & \frac{1}{4} \end{pmatrix}$$

If the probabilities along each row are not the same then you must use a higher power.

Thus, in the long run, there is a $\frac{1}{8}$ chance of being at A, a $\frac{3}{8}$ chance of being at B, and so on. The start vertex is irrelevant. The probabilities, of course, must add to give 1.

Transition Matrices – Practical Application: Consider a website where individual pages have links to other pages. If we treat the links as a network we can draw a graph showing the structure of the website.

For example, let's take a simple website with just five pages, and where we use directed edges to show links between the pages.

Assuming that links on a page are clicked on with equal probability, the transition matrix is:

$$T = \begin{pmatrix} 0 & \frac{1}{2} & 0 & 0 & 1 \\ \frac{1}{2} & 0 & \frac{1}{3} & 0 & 0 \\ \frac{1}{2} & \frac{1}{2} & 0 & 0 & 0 \\ 0 & 0 & \frac{1}{3} & 0 & 0 \\ 0 & 0 & \frac{1}{3} & 1 & 0 \end{pmatrix}$$

Taking a high power of T leads to the following steady state probabilities for landing on any page:

Page number	1	2	3	4	5
Probability	$\frac{5}{18}$	$\frac{2}{9}$	$\frac{1}{4}$	$\frac{1}{12}$	$\frac{1}{6}$

This gives us a simple measure of page rank, since more people will land on page 1 in the long run than any other page. The actual ranking is: 1, 3, 2, 5, 4.

3.15 Graph Algorithms

Having constructed a graph to represent a real-world situation, we need methods, other than trial and error, for solving problems such as shortest route, least cost and so on. You need to be able to remember the algorithms, as well as which problem each one solves!

Minimum spanning tree: Given a weighted connected graph, a common problem is to find the most efficient way to ensure that all the vertices are connected to all the others (not necessarily directly). Such a simplified graph is called a *spanning tree*: it must be a tree because any cycles would include redundant edges.

Any spanning tree for a graph with n vertices will have $n - 1$ edges.

In the graphs below, T_1 and T_2 are each spanning trees for G:

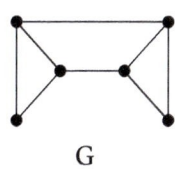

G

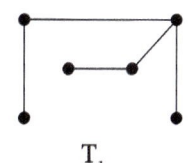

T_1

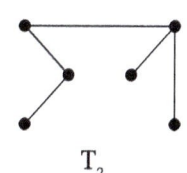

T_2

The spanning tree with the lowest total cost is called the *minimum spanning tree* (or *minimum connector*). There are two algorithms for finding minimum spanning trees: Kruskal's and Prim's.

3.16 Kruskal's Algorithm

1. Select the smallest edge in the graph. This becomes the first edge in your solution.

2. Select the smallest unused edge in the original graph and, as long as it does not form a cycle, add it to your solution. At this stage, it does not have to be connected.

3. Repeat step 2 until all vertices are connected.

Note: if you have a choice of edges at any stage, select one arbitrarily.

Example: The graph shows the hubs of 8 computer networks and the costs of connecting the hubs. It is required to connect all the hubs at minimum cost. Use Kruskal's algorithm to find the minimum spanning tree, showing each stage of the working.

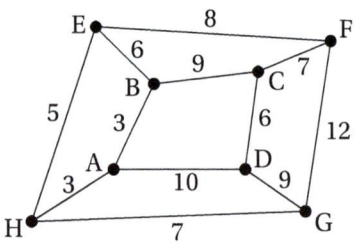

Solution:

Edge	Cost	Choice	Notes
AB	3	1st	Could have chosen AH
AH	3	2nd	
HE	5	3rd	
CD	6	4th	BE would have created a cycle
CF	7	5th	Or GH
HG	7	6th	
EF	8	7th	Stop at 7 edges

Build up the tree yourself to watch it taking shape. The final result looks like this:

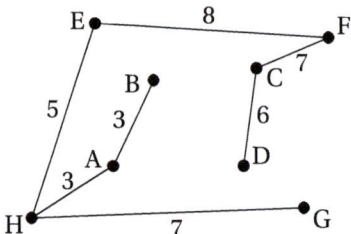

Kruskal's algorithm is fine for small networks, but with larger networks it becomes harder to identify cycles – especially if the problem is to be solved with a computer program.

3.17 Prim's algorithm

Prim's algorithm takes a more systematic approach than Kruskal's:

1.   Select any vertex as a start vertex. This is the first part of your solution.

2.   Consider all the edges which connect to any vertex already in your solution to one not in the solution and choose the smallest. Add it to your solution.

If there is more than one choice, choose arbitrarily.

3.   Repeat step 2 until all the vertices have been added.

Let's see how this works with the example on the previous page. We shall start arbitrarily at vertex F. In each case the most recent edge is shown in bold.

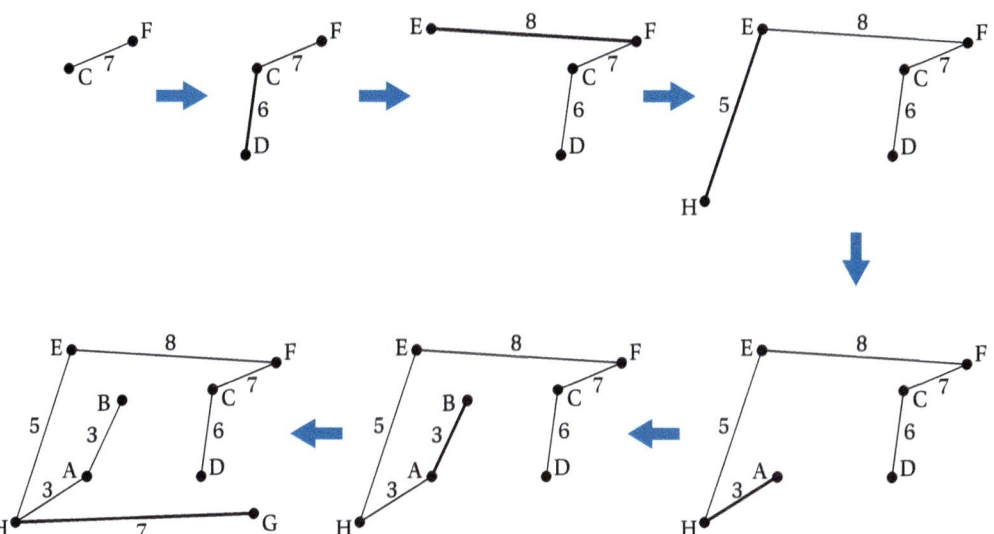

Note that at the final stage edge BE (weight 3) was cheaper that HG, but would not have introduced a new vertex. Now try Prim's algorithm for yourself starting at any other vertex – you will always end up with the same minimum spanning tree.

Prim's algorithm with a matrix: If the weighted adjacency matrix for a network is drawn we can adapt Prim's algorithm to find the minimum spanning tree. This means that we do not need to draw the network – particularly useful if it is very large.

1.   Select any vertex as the start vertex. Select its column by circling its column header and delete its row.

2.   Find the smallest value in any selected column and non-deleted row. Add the corresponding vertex to the solution, circle its header and delete its row.

3.   Repeat step 2 until all columns have been selected.

This is easier to see in practice than it is to explain, so let's see it in action with the same problem as above, by drawing the weighted graph and then starting at vertex F.

	A	B	C	D	E	F	G	H
A		3		10				3
B	3		9		6			
C		9		6		7		
D	10		6				9	
E		6				8		5
F			7		8		12	
G				9		12		7
H	3				5		7	

7 is smallest – C is the new vertex.

	A	B	C	D	E	F	G	H
A		3		10				3
B	3		9		6			
C		9		6		7		
D	10		6				9	
E		6				8		5
F			7		8		12	
G				9		12		7
H	3				5		7	

6 is smallest – D is the new vertex.

	A	B	C	D	E	F	G	H
A		3		10				3
B	3		9		6			
C		9		6		7		
D	10		6				9	
E		6				8		5
F			7		8		12	
G				9		12		7
H	3				5		7	

8 is smallest – E is the new vertex.

	A	B	C	D	E	F	G	H
A		3		10				3
B	3		9		6			
C		9		6		7		
D	10		6				9	
E		6				8		5
F			7		8		12	
G				9		12		7
H	3				5		7	

5 is smallest – H is the new vertex.

	A	B	C	D	E	F	G	H
A		3		10				3
B	3		9		6			
C		9		6		7		
D	10		6				9	
E		6				8		5
F			7		8		12	
G				9		12		7
H	3				5		7	

3 is smallest – A is the new vertex.

	A	B	C	D	E	F	G	H
A		3		10				3
B	3		9		6			
C		9		6		7		
D	10		6				9	
E		6				8		5
F			7		8		12	
G				9		12		7
H	3				5		7	

3 is smallest – B is the new vertex.

	A	B	C	D	E	F	G	H
A		3		10				3
B	3		9		6			
C		9		6		7		
D	10		6				9	
E		6				8		5
F			7		8		12	
G				9		12		7
H	3				5		7	

7 is smallest – G is the new vertex.

	A	B	C	D	E	F	G	H
A		3		10				3
B	3		9		6			
C		9		6		7		
D	10		6				9	
E		6				8		5
F			7		8		12	
G				9		12		7
H	3				5		7	

Solution.

At each stage, the smallest edge has been boxed – this allows us to keep a record of which edges have been used, and hence draw the minimum spanning tree if necessary. Look at the final table – each boxed number indicates an edge contained in the solution.

The diagram shows the cost of connecting locations in a small town with fibre to create a fast broadband network.

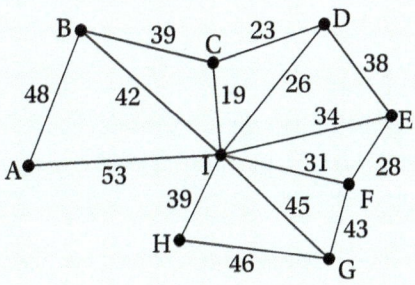

(a) Use Kruskal's algorithm to find a minimum spanning tree. List the arcs in the order you add them, with each of their weights.

(b) Starting at A, use the table method for Prim's algorithm to find a minimum spanning tree, stating the order in which arcs are added to the tree.

(c) (i) Draw a minimum spanning tree.

   (ii) State the total cost for the tree in (i).

(d) State a disadvantage of creating such a fibre network using a minimum spanning tree.

A new spanning tree is to be created which must include arcs DI and HG, and which has the lowest total cost.

(e) Explain which algorithm you would use to complete the tree, and how it should be adapted.

(c) (i)		A full answer to this question can be found on the website.

(c) (i)
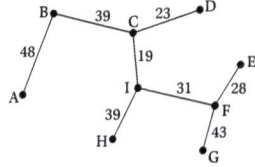

   (ii) Minimum spanning tree (cost ≠ 270)

(d) No redundancy. If one link is faulty, all the locations beyond will lose access.

(e) Kruskal. Start with DI and HG, and then use the algorithm to complete the tree.

3.18 Chinese Postman Problem

The Chinese postman problem (so called because it was devised by a Chinese mathematician) is that of finding the shortest route around a weighted graph such that every edge is traversed. This is equivalent to walking every street in a town. Note that edges may need to be traversed more than once, and vertices visited several times. The "postman" will start and finish at the same point, so effectively we are looking to create an Eulerian circuit.

The algorithm for solving the problem varies according to the number of odd vertices: in this course you will only need to deal with graphs with 0, 2 or 4 odd vertices.

No odd vertices: We saw on page 98 that a graph with all even vertices will be traversable. Thus there must be a route which starts and ends at a given vertex, and uses each edge just once. The distance can be found simply by adding up the total weight of all the edges.

2 odd vertices: Eulerian circuits will only exist if there are no odd vertices; so all we have to do is to join the two odd vertices along the shortest route, and they become even. The total distance (or weight) will now be the sum of all the distances plus the one extra edge distance.

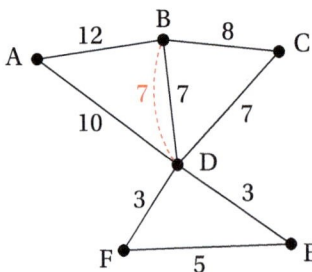

This graph has odd vertices B and D. If we join them (orange line), all the vertices will be even. (If the edges represent roads, then adding extra edge BD is equivalent to walking along the road both ways).

The total distance is now $55 + 7 = 62$, and a possible circuit starting at A, say is ABCDBDEFA.

You will be able to find a circuit starting at **any** vertex.

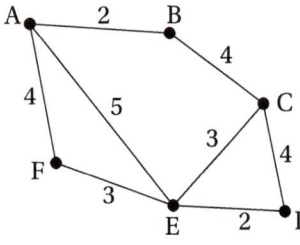

In the previous example, there was a single edge between the two odd vertices. In this graph, the two odd vertices are A and C, and the shortest route between them is ABC. So this time we would need to double up edges AB and BC (we can't join A directly to C without building a new road)! Check you can find a circuit if you do that, and the shortest distance is 33.

4 odd vertices: Once again we need to add edges to the odd vertices to make them even. But which do we join to which? Here is the algorithm:

1. List the four odd vertices.

2. List the possible pairings of the odd vertices.

3. For each pairing, find the edges that connect the vertices with minimum weight.

4. Find the pairings where the sum of the weights is a minimum.

5. On the original graph, add the edges that have been found in step 4.

6. The length of the optimal circuit is the sum of weights of the original edges + edges added in step 5.

Situations such as these are generically known as "the route inspection problem."

Example: The diagram shows a network of roads in a village, with distances in metres. Each road is to be inspected for potholes, starting at the crossroads at G.

Devise a route for the inspector which includes each road, but ensures he travels the shortest possible distance.

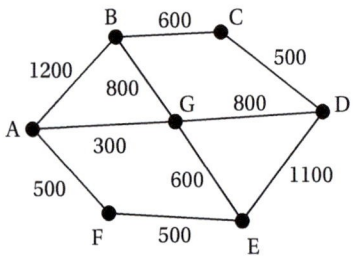

Solution: We shall use the Chinese postman algorithm.

- The odd nodes are A, B, D, E.

- Possible pairings are: AB and DE

  AD and BE

  AE and BD

- Minimum distances for the pairs: AB: 1100, DE: 1100. Total: 2200

  AD: 1100, BE: 1400. Total: 2500

  AE: 900, BD: 1100. Total: 2000

Note that, although there is a direct connection from A to B, it is not the shortest route.

- AE and BD provide the minimum total. Add edges AG, GE; BC, CD.

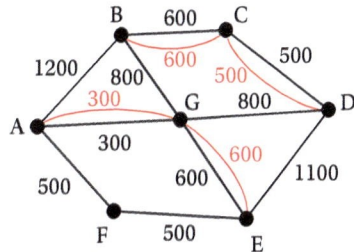

- Optimal route has length 6900 + 2000 = 8900 m
- Possible inspection route: GBAFEGEDCBCDGAG.

Start and end at different vertices: This version of the Chinese postman algorithm is useful for the postman who wants to start at the post office, but end up somewhere else – perhaps his home! We need a graph which is semi-traversable: it will have 2 odd vertices. If it already has 2 odd vertices, then the trail will start at one of them and finish at the other. If it has 4 odd vertices, then we need to start at one, end at another, and pair up the remaining two. For example, in the graph above, suppose we want to start at B and end at D. We leave the extra edges AG and GE, and you should then be able to find a trail.

eg: BCDGBAGAFEGED

If either the start or end vertices are even then they must be made odd, and other adjustments made so that the graph has two odd vertices.

Example: Find a semi-Eulerian trail, starting at A and ending at C, in the following graph:

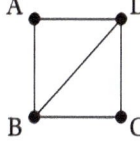

Solution: A and C are even, so an edge can be drawn between them to make them odd. There would then be four odd vertices, so another edge is drawn between B and D to make them even.

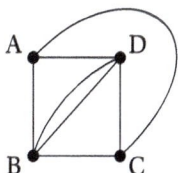

A trail can now be found by inspection, eg: ADCDCBAC.

3.19 Travelling Salesman Problem (TSP)

The route inspector needs to travel down each road (edge), the salesman needs to visit each town (vertex) and end up back where he started – in other words, we need to find a Hamiltonian cycle (see page 97) of minimum weight. However, not all graphs have Hamiltonian cycles and this makes the travelling salesman problem more complex than it might at first sight appear.

The number of Hamiltonian cycles for K_n is $\frac{1}{2}(n-1)!$, so for K_6 there will be 60 possible cycles.

All **complete** graphs have Hamiltonian cycles but there is no known way of finding which cycle has the least weight. Since the number of cycles can be extremely high, simply checking each one is a very inefficient way to find a solution. Instead, methods focus on finding not the optimal solution, but simply a reasonably good one. In particular, there are algorithms for finding the lower and upper bounds.

Practical and Classical TSP: In practice, the travelling salesman will not mind revisiting a town if the overall route is the shortest – and he may **have** to do this if the graph is not complete (ie there may not be a Hamiltonian cycle). This is known as the "practical problem". Since all complete graphs do have Hamiltonian cycles, then solving the TSP on a complete graph is known as the "classical problem", with the additional criterion that the triangle inequality holds. This means that, for any triangle of edges in the graph, no single side can be longer than the sum of the other two sides.

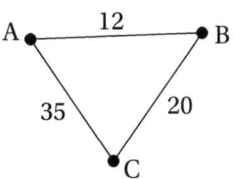

Triangle inequality: To get from A to C, it is quicker to go via B. In the classical problem, 35 will be replaced by 32.

Lower bound: This algorithm will find a value which must be less than the optimal solution (for any weighted graph):

1. Choose an arbitrary vertex, say X. Find the sum of the two smallest weights of all the edges which touch X.

2. Remove X from the network, and all the edges which touch X. Find the weight of the minimum spanning tree for this network.

3. The sum of the totals in steps 1 and 2 is a lower bound.

Consider the weighted graph G_1 shown in the diagram.

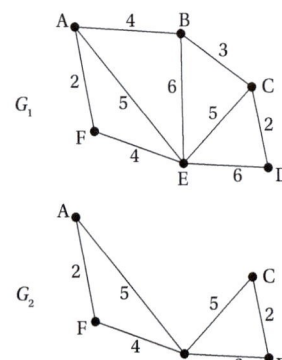

1. Choose B. The two smallest weights add to give 7.

2. Remove B and all its incident edges - we get G_2. The minimum spanning tree for this network is formed by edges AF, FE, EC and DC and its total weight is 13.

3. Thus, $7 + 13 = 20$ is a lower bound.

We might be able to increase this lower bound by removing each of the vertices in turn and carrying out the same process. Try this yourself, and see if you end up with the following table of lower bounds:

Remove vertex	A	B	C	D	E	F
Lower bound	20	21	21	21	20	20

Thus we now know that the optimal solution will have a weight of at least 21.

Converting the practical to the classical problem: Since the classical TSP is a complete graph with shortest distances between every pair of vertices, we need to do two things when converting:

*   Add edges where vertices are not directly connected. The weight of such edges is the shortest indirect route between the vertices.

*   Check for pairs of vertices where the triangle inequality doesn't hold, and replace the current weight with the shorter one.

Once this has been done, any solutions for the classical problem will be exactly the same for the converted practical problem.

The conversion can be done either by creating a new graph or by drawing up a least distances table.

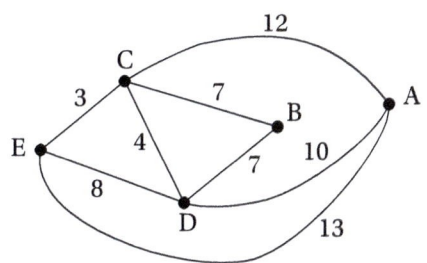

Consider this graph. Which vertices are not directly connected? AB, BE. What are the minimum distances between them? AB = 17, BE = 10.

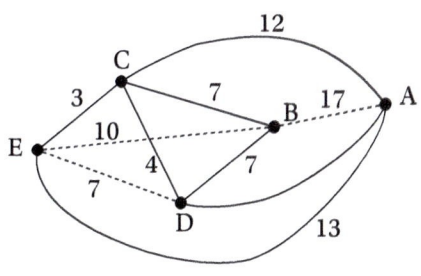

Now look at every triangle – do they all obey the triangle inequality. All of them do except CDE because DE > CE + CD. So we correct this by relabelling the weight on DE as 7. We can now draw the new graph – additions shown as dotted lines.

> Note that we have to draw BE crossing CD. But this doesn't create a new vertex.

The least distances table looks like this:

	A	B	C	D	E
A		17	12	10	13
B	17		7	7	10
C	12	7		4	3
D	10	7	4		7
E	13	10	3	7	

Upper bound: The upper bound for the TSP can be found using the *nearest neighbour algorithm*. Note that this requires a classical TSP, so if the question involves a practical TSP you must first convert it.

Here's the algorithm:

1. Call the start point the "current vertex"
2. Find the nearest unvisited vertex to the current vertex, and move to that vertex. This becomes the new current vertex.
3. Repeat step 2 until all vertices have been visited
4. Return directly to the start vertex.

> This algorithm should not be confused with Prim's which finds the nearest vertex to any visited vertex, not just the current one.

By adding all the weights of the edges you have moved along you will have an upper bound for the problem. The main problem with this method is that we have no control over the final edge, which may be very long.

Let's use the least distances table in the previous section to find an upper bound for the classical TSP graph. As with Prim's algorithm, you can circle each vertex you visit and then cross out its row. Starting at A, we get the following circuit:

AD = 10, DC = 4, CE = 3, EB = 10, BA = 17. Total weight = 44.

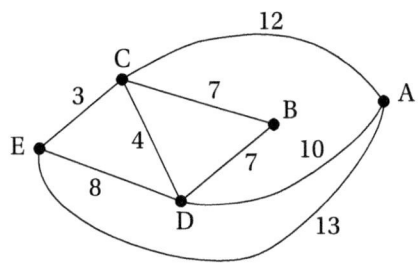

As an exercise, take the same network (reproduced here) and use each vertex as a start point to find possible lower bounds, and then again, using the nearest neighbour algorithm, to find the possible upper bounds. Hence state the least range within which the optimal solution lies.

Answers	Lower Bound:	Remove A:	$10 + 12 + 14 = 36$
		Remove B:	$7 + 7 + 17 = 31$
		Remove C:	$3 + 4 + 25 = 32$
		Remove D:	$4 + 7 + 22 = 33$
		Remove E:	$3 + 8 + 21 = 32$
	Upper Bound:	Start at A:	ADBCEA = 40
			ADECBA = 44
		Start at B:	BCEDAB = 44
			BDCEAB = 44
		Start at C:	CEDBAC = 46
		Start at D:	DCEBAD = 44
		Start at E:	ECDBAE = 44

Thus the optimal solution lies between 36 and 40.

The diagram shows the network of paths connecting five statues in a public park, and the walking time, in seconds, along each path. The park administrator wants to devise a walking route which visits each statue in the least time so that they can be inspected for damage.

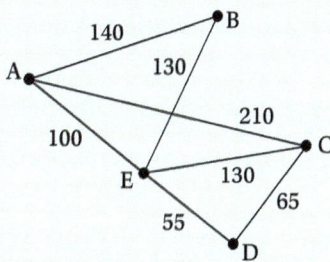

(a) What is the standard name for this type of problem?

(b) (i) Use an algorithm which begins by deleting statue A from the network to find a possible value for the lower bound of the optimal walking time. You must explain how the value is calculated.

 (ii) The same algorithm is used with statue D being first deleted, and this leads to a lower bound of 480 s. State which of the two values you would now use as the lower bound.

(c) Draw up a table of least distances between each of the five statues.

(d) Starting at statue A, use the nearest neighbour algorithm to calculate a possible upper bound.

(e) Starting at A, suggest a route whose time lies between the lower and upper bounds.

(a) Travelling salesman problem.

(b) (i) Two shortest edges from B are 140 + 100 = 240.

 This leaves the following minimum spanning tree.

 Lower bound is 240 + 250 = 490.

 (ii) 490

(c)

	A	B	C	D	E
A	–	140	210	155	100
B	140	–	250	185	130
C	210	250	–	65	120
D	155	185	65	–	55
E	100	130	120	55	–

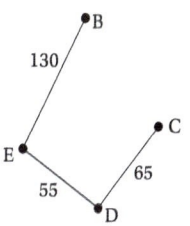

(d) A – E – D – C – B – A = 100 + 55 + 65 + 250 + 140 = 610

(e) For example, ACDEBA = 600.

Geometry and Trigonometry: Long Answer Questions

Starting on the next page is a selection of Paper 2 style exam questions. The answers are given here, but full working may be found on the Peak Study Resources website.

1. There are three refuge huts set up in a rectangular area of a National Park. The positions of the huts are shown on the diagram with a scale in kilometres. The huts have the following coordinates: A (60, 50), B (10, 20) and C (90, 20). The boundary of the National Park is formed by the axes and the lines $x = 100$ and $y = 60$.

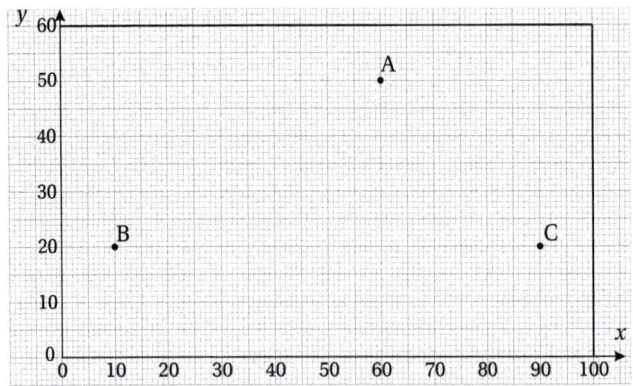

(a) Write down the equation of the perpendicular bisector of [BC]

(b) Find the equation of the perpendicular bisector of [AC]

In bad weather people are advised to go to the refuge that is closest to them.

Given the perpendicular bisector of [AB] has equation $5x + 3y = 280$:

(c) Draw a Voronoi diagram showing the areas that are closest to each of the huts indicating clearly the point of intersection of the three perpendicular bisectors.

(d) Find the area of the region in the National Park in which people would go to

   (i)   Hut A

   (ii) Hut C

   (iii) Hut B

(e) A fourth refuge hut stand is to be added inside the boundary of the National Park, at a point as far as possible away from all the other three huts.

   (i)   State the coordinates of the position at which it should be built

   (ii) How far will it be from the other three huts?

Answers:

(a) $x = 50$

(b) $y = x - 40$

(c)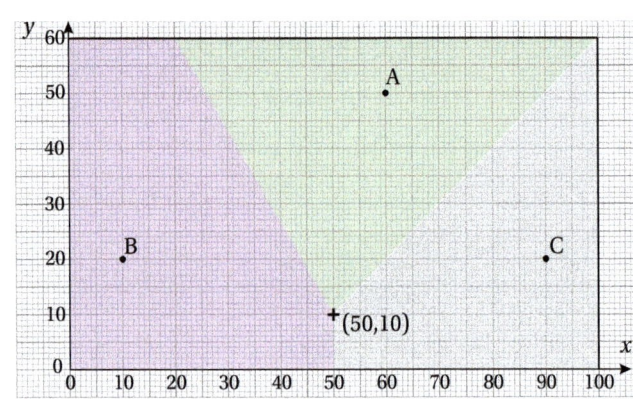

(d) (i)   2000 km²

     (ii) 1750 km²

     (iii) 2250 km²

(e) (i)   (50, 10)

     (ii) 41.2 km

2. A cone has base diameter DE = 10 cm and height 16 cm.

    (a) Calculate the slant height AD of the cone.

    (b) Calculate the volume of the cone.

    A *frustum* is formed by cutting off a smaller cone from the top of the original cone. The frustum has height 8 cm and the smaller cone has diameter BC = 5 cm.

    (c) Show that the volume of the frustum is 366.5 cm³.

    A cylinder has the same height and volume as the frustum.

    (d) Calculate the radius of the cylinder.

    (e) Calculate the volume of a hemisphere of radius 1.5 cm

    The hemisphere is repeatedly filled with water and emptied into the cylinder.

    (f) (i) How many complete hemispheres of water can be emptied into the cylinder without the cylinder overflowing.

        (ii) Calculate the remaining height in the cylinder above the water level.

Answers:

    (a) 16.8 cm

    (b) 418.9 cm³

    (d) 3.82 cm

    (e) 7.07 cm³

    (f) (i) 51

        (ii) 0.131 cm³

3. Scientists investigating artificial intelligence create a maze consisting of a series of small rooms as shown below:

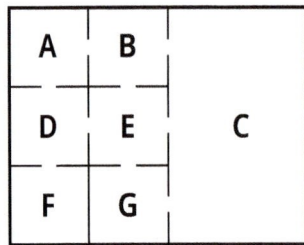

(a) Draw the maze as a graph in which the rooms are the vertices and the openings are the edges.

(b) (i) State whether or not the graph has a Hamiltonian cycle, justifying your answer.

(ii) Write down which openings would need to be added to the rooms so that the graph would have an Euler circuit. There can be more than one opening in a single wall.

The scientists put a small robot into the maze. When in a room the robot is equally likely to leave by any of the openings. The scientists hope that when the robot has been left there a long time it will have learned the maze structure.

(c) Construct a transition matrix to describe the robot's movements.

(d) Find the room or rooms in which the robot spends

(i) the most time

(ii) the least time

In each case give the long term proportion of time spent in these rooms.

_____

Answers:

(a) See diagram

(b) (i) No, because it would be necessary to visit D twice to get to F.

(ii) F to D, B to C.

(c)
$$\begin{pmatrix} 0 & \frac{1}{3} & 0 & \frac{1}{3} & 0 & 0 & 0 \\ \frac{1}{2} & 0 & \frac{1}{3} & 0 & \frac{1}{4} & 0 & 0 \\ 0 & \frac{1}{3} & 0 & 0 & \frac{1}{4} & 0 & \frac{1}{2} \\ \frac{1}{2} & 0 & 0 & 0 & \frac{1}{4} & 1 & 0 \\ 0 & \frac{1}{3} & \frac{1}{3} & \frac{1}{3} & 0 & 0 & \frac{1}{2} \\ 0 & 0 & 0 & \frac{1}{3} & 0 & 0 & 0 \\ 0 & 0 & \frac{1}{3} & 0 & \frac{1}{4} & 0 & 0 \end{pmatrix}$$

(d) (i) E $0.222 = \frac{2}{9}$

(ii) F $0.0556 = \frac{1}{18}$

4. A sequence of right-angled, isosceles triangles (T_n) are shown below. T_n is formed from T_{n-1} by rotating the triangle 45° clockwise about (0,0) and enlarging by a factor $\frac{1}{\sqrt{2}}$, centre (0,0). T_1 is the triangle OAB shown with vertices at (−4, 0), (−4, 4) and (0, 0).

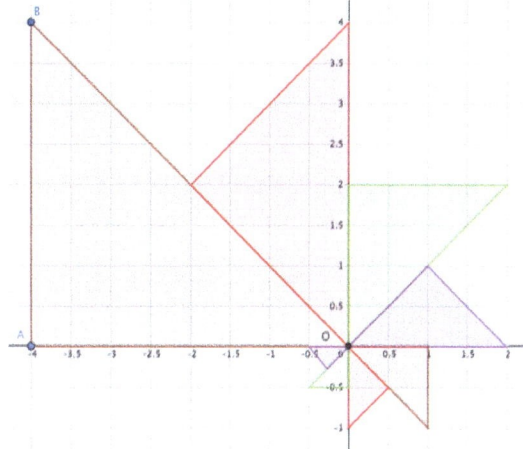

Let A_n be the area of triangle T_n.

(a) (i) Write down A_1

    (ii) Given that the areas of successive triangles form a geometric progression, find an expression for A_n.

    (iii) Hence find the total area for the eight triangles.

Let E be the enlargement matrix and R the rotation matrix that describe the transformation T_{n-1} to T_n.

(b) Write down:

    (i) E (ii) R

If (x_1, y_1) is a point in T_1 and (x_n, y_n) is a point in T_n then $\begin{pmatrix} x_n \\ y_n \end{pmatrix} = M^{n-1} \begin{pmatrix} x_1 \\ y_1 \end{pmatrix}$.

(c) (i) Find M as a single 2×2 matrix

    (ii) **Hence** calculate the image of point A in T_8.

_____

Answers:

(a) (i) 8

    (ii) $8 \times \left(\frac{1}{2}\right)^{n-1}$

    (iii) 15.9375

(b) $E = \begin{pmatrix} \frac{1}{\sqrt{2}} & 0 \\ 0 & \frac{1}{\sqrt{2}} \end{pmatrix} = \begin{pmatrix} 0.707 & 0 \\ 0 & 0.707 \end{pmatrix}$ $R = \begin{pmatrix} \frac{1}{\sqrt{2}} & \frac{1}{\sqrt{2}} \\ -\frac{1}{\sqrt{2}} & \frac{1}{\sqrt{2}} \end{pmatrix} = \begin{pmatrix} 0.707 & 0.707 \\ -0.707 & 0.707 \end{pmatrix}$

(c) (i) $M = \begin{pmatrix} 0.5 & 0.5 \\ -0.5 & 0.5 \end{pmatrix}$

    (ii) (−0.25, −0.25)

5. **Part A**

The table below shows the distances (in m) between all the blocks of a university campus. An internal telephone network is to be installed linking these blocks. Use Prim's algorithm on a matrix, starting at A, to find the minimum length of the cable.

	A	B	C	D	E	F
A	–	700	800	200	600	700
B	700	–	250	600	200	100
C	800	250	–	700	300	350
D	200	600	700	–	600	650
E	600	200	300	600	–	100
F	700	100	350	650	100	–

Part B

Consider the travelling salesman problem for the following network:

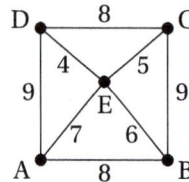

(a) Find the lower bound when each of the following vertices is removed:

(i) A

(ii) E

(b) Add two weighted edges to convert the graph from a practical to a classical Travelling Salesman Problem.

(c) Use the nearest neighbour algorithm, starting at vertex A, to find an upper bound for the Travelling Salesman Problem.

(d) Find a solution with length 35.

_____

Answers:

Part A

1250m

Part B

(a) (i) 30

    (ii) 34

(b) BD = 10, AC = 12

(c) 36

(d) A – D – E – C – B – A

6.   Consider the following adjacency matrix (M)

$$\begin{array}{c}\\A\\B\\C\\D\\E\\F\\G\end{array}\begin{array}{c}A\ B\ C\ D\ E\ F\ G\\\left(\begin{array}{ccccccc}0&0&1&0&0&1&1\\0&0&0&0&1&0&0\\1&0&0&0&0&0&1\\0&0&0&0&1&0&0\\0&1&0&1&0&0&0\\1&0&0&0&0&0&1\\1&0&1&0&0&1&0\end{array}\right)\end{array}$$

(a)   What features of the adjacency matrix tells you the graph is

   (i)   simple

   (ii)   undirected

(b)   Explain how the matrix can be used to calculate the order of each vertex

   The matrix (M) represents connecting flights between cities

(c)   (i)   Find M^2

   (ii)   Hence find the city, other than B itself, that can be reached from B using exactly two flights.

(d)   (i)   By considering a sum of powers of M prove that the graph is not connected, fully justifying your answer.

   (ii)   Write down the two sets of connected cities.

An extra flight between E and F, returning F to E, is now added.

(e)   Find which pairs of cities cannot be connected in two or fewer flights.

_____

Answers:

(a)   (i)   Zeroes in leading diagonals; only entries are ones and zeroes.

   (ii)   Symmetrical around leading diagonal

(b)   Row or column totals

(c)   (i)

$$\begin{array}{c}\\A\\B\\C\\D\\E\\F\\G\end{array}\begin{array}{c}A\ B\ C\ D\ E\ F\ G\\\left(\begin{array}{ccccccc}3&0&1&0&0&1&2\\0&1&0&1&0&0&0\\1&0&2&0&0&2&1\\0&1&0&1&0&0&0\\0&0&0&0&2&0&0\\1&0&2&0&0&2&1\\2&0&1&0&0&1&3\end{array}\right)\end{array}$$

   (ii)   D

(d)   (i)   Calculate $M + M^2 + ... + M^6$

   If connected, there would be at least one route between every pair of cities. However, A to B for example shows no connection.

   (ii)   A, C, F, G.   B, D, E.

(e)   From $M + M^2$, the following pairs have no connections:

   AB, AD, BC, BG, CD, CE, DG.

117

7. Consider the three points A(–3, 4, 2), B(2, 4, 0) and C(5, 0, –2)

   (a) Write down the coordinates of M, the midpoint of AB.

   Florence is creating a 3-d model as part of a design project. She first fixes a long metal rod R_1 which passes through the origin and the point M.

   (b) Find a vector equation for the line created by rod R_1.

   A second rod R_2 passes through C and is parallel to vector $\begin{pmatrix} a \\ 4 \\ 2 \end{pmatrix}$. R_2 intersects R_1 at a point N.

   (c) (i) Show that a = –3

       (ii) Find N.

   Florence now fixes a piece of card with corners OABC.

   (d) Prove that OABC is a rhombus, but not a square.

   Finally Florence fixes a triangular piece of card with corners OAM.

   (e) (i) Calculate $\overrightarrow{OA} \times \overrightarrow{AM}$

       (ii) Hence or otherwise show that the area of triangle OAM is $\sqrt{30}$.

   (f) Prove that the area of OABC is four times the area of OAM.

   _____

   Answers:

   (a) M = (–0.5, 4, 1)

   (b) $r = t\begin{pmatrix} -0.5 \\ 4 \\ 1 \end{pmatrix}$ (or any multiple).

   (c) (i) a = –3

       (ii) N = (–1, 8, 2)

   (d) For example, prove opposite sides are equal vectors, and one angle ≠ 90°.

   (e) (i) $\begin{pmatrix} -4 \\ 2 \\ 10 \end{pmatrix}$

Chapter 4: STATISTICS AND PROBABILITY

4.1 Definitions

A *population* is a set from which statistics are drawn. A *sample* is a subset drawn from the population. In a random sample, every member of the population is equally likely to be chosen. There are several different *sampling techniques*, each with advantages and disadvantages. A sampling technique may introduce *bias* – for example, selecting a sample of people in a shopping centre at 11am will not include those who are at work.

Sample statistics (such as the mean) can be used to estimate population statistics. *Discrete* data are restricted to certain values only (often integers) whereas *continuous* data can take any values. The *frequency* is the number of times a particular value occurs. When collecting data, some items may appear to be very extreme when compared to the rest of the data. Such items are called *outliers* and consideration must be given to whether they *could* be valid, or whether they are incorrect; and also how to deal with them.

Numerical data is usually collected into a *frequency table* and can then be split into *groups* or *classes*. The *boundaries* of the classes must be dealt with carefully, especially for continuous data. Consider a table of weights which begins like this:

Weight (kg)	Frequency
0–10	4
10–20	12
20–30	18

Into which class would a weight of 10kg be put? It would be better if the first group were labelled $0 \leq w < 10$ and the second $10 \leq w < 20$, then 10 would fall into the second group. The *interval width* in this case is 10, and the *mid-interval value* of the first group is 5 and so on. Data can be appreciated more when displayed in a diagram and the *frequency histogram* is the simplest way to display grouped data. A frequency histogram (often called a *bar chart*) uses equal class intervals.

4.2 Averages

Properly called "measures of central tendency", there are three types of average you need to know: mean, median and mode. An average is a single statistic which can be used to represent a whole group, although this isn't true of the mode which merely tells us the "most popular" value.

Examples of populations:
People who live in Europe
People who drive
Apples grown in France
Cars made in 2002

Examples of discrete data:
Shoe sizes
Goals scored by a team
Number of chocolates in a box

Examples of continuous data:
Weights of people
Athletes' times to run 100m
Heights of mountains

The mean: To calculate the mean, add all the numbers together and divide by the number of values, n. So mean $= \dfrac{\sum x_i}{n}$, where the separate values are x_1, x_2, x_3 and so on. The symbol for sample mean is $\bar{x}$. Note that $n\bar{x} = \sum x_i$.

Example: In 9 games I have scored a mean of 12.8 points. In the 10th game I score 16 points – what is my new mean?

Solution: The total score in the first 9 games is $9 \times 12.8 = 115.2$. My new total in 10 games is $115.2 + 16 = 131.2$, so my new mean is $\dfrac{131.2}{10} = 13.12$.

Here's a similar question:

> 100 people are staying at a hotel: 68 are men and 32 women. The men have a mean height of 1.75 m and the women have a mean height of 1.64 m. Find the mean height of the 100 people.

The answer is 1.71 m.

If you can't get there, you'll find full working on the website.

If the data is in a frequency table – such as the one below showing how many pupils were absent during a month – then the total value is calculated by multiplying each value by its frequency and summing the results.

Pupils absent (x)	No of days (f)	fx
0	20	0
1	4	4
2	3	6
3	3	9
TOTAL	**30**	**19**

There were a total of 19 days absence over a period of 30 days. So the mean number of absences per day was $\dfrac{19}{30} = 0.63$. (It is a common mistake to divide 19 by 4, the number of classes).

If the data is presented in a *grouped frequency table*, the same procedure is followed except that the mid-interval value of each group is used to represent the x value for each group. This means that the **actual** data values are unknown and in this case the mean is only an estimate.

Always check if the answer is "reasonable." Look at the distribution of weights – does 30.7 look like the mean?

Weight of apples (w)	No of apples (f)	Mid interval	fx
$20 \le w < 25$	12	22.5	270
$25 \le w < 30$	20	27.5	550
$30 \le w < 35$	25	32.5	812.5
$35 \le w < 40$	17	37.5	637.5
TOTAL	**74**		**2270**

Estimated mean weight of an apple is $\dfrac{2270}{74} = 30.7$ g.

The table shows the scores of competitors in a competition.

Score	10	20	30	40	50
Number of competitors with this score	1	2	5	k	3

The mean score is 34. Find the value of k.

Total score = $10 + 40 + 150 + 40k + 150 = 350 + 40k$

Number of competitors = $1 + 2 + 5 + k + 3 = 11 + k$

Mean is $\dfrac{350 + 40k}{11 + k} = 34$

$350 + 40k = 34(11 + k)$

$350 + 40k = 374 + 34k$

$6k = 24$

$k = 4$

At first sight you may be puzzled as to how to tackle a question like this. But, knowing the formula for the mean, just carry on and see what happens!

If you find the algebra too much, you can of course use the equation-solving functionality of your GDC.

Your GDC will calculate the mean of a set of numbers within its statistical functions. But be careful: if you want the mean of a frequency table you will need two lists (the values and the frequencies), and the GDC will need to know that the second list contains the frequencies. If you're uncertain about this, try it with the table above.

Median: If a set of values is listed in order, the middle value is the median. It is another type of average: there are as many values above the median as below it. Unlike the mean, it is unaffected by particularly large or small values. In the following list there are 15 values so the 8th is the middle one (7 below it, 7 above it).

In general, if there are n values, the median is in the $\dfrac{n+1}{2}$th position.

$$1\ 1\ 3\ 5\ 6\ 6\ 6\ \underline{7}\ 7\ 9\ 10\ 10\ 12\ 15\ 18 \rightarrow \text{median} = 7$$

If there is an even number of values, find the mean of the middle two to calculate the median.

$$24\ 26\ 27\ \underline{27\ 29}\ 30\ 30\ 33 \rightarrow \text{median} = 28$$

If the data is in the form of a frequency table, then the calculation depends on whether it is discrete or continuous.

Discrete distribution

x	1	2	3	4	5	6
f	4	11	17	25	14	4

There are 75 values, so the median will be the 38th. The first 4 values are 1s, the next 11 are 2s, making 15 values so far. Another 17 are 3s making 32 values. So the 38th value must be in the next box, and thus the median is 4.

Beware! If you enter a grouped frequency table, you will not get correct values for the median and the quartiles.

Continuous distribution

x	0 –	5 –	10 –	15 –	20 –	25 – 30
f	4	11	17	25	14	4

This time, the values are spread throughout each class, so the 38th value will be the 6th in the class 15–20. Interpolating, median $= 15 + \dfrac{6}{25} \times 5 = 16.2$

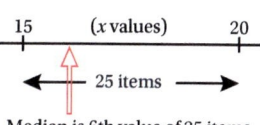

Median is 6th value of 25 items

Cumulative frequency tables: It is slightly easier to estimate the median from a frequency table if it is first converted into a cumulative frequency table. Whether the data is discrete or continuous, the method is the same. Each value of cumulative frequency measures how many x values there are in total up to that point. The two tables above convert into the following:

x	≤ 1	≤ 2	≤ 3	≤ 4	≤ 5	≤ 6
Cumul. f	4	15	32	57	71	75

In the first table we can see that there are 32 values up to 3, so the 38th value must be contained in the next group and is 4.

x	< 5	< 10	< 15	< 20	< 25	< 30
Cumul. f	4	15	32	57	71	75

Note that in the conversion of the grouped frequency table, the "up to" points are the top of each group.

In the second table we have to recalculate the fact that there are 25 values in the group 15–20, and then go on to the calculation shown above. The advantage here is not so great, but we can go one stage further and draw a cumulative frequency graph to help us.

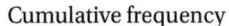

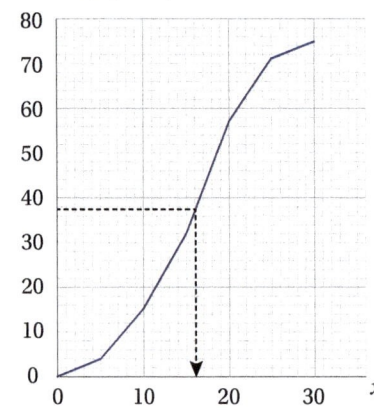

The points in the table are plotted and are joined either by straight lines or a smooth curve. To find the median, a line is drawn to the right from 37.5 (the middle value of the distribution) and down to the x axis.

The median can be seen to be about 16.

Quartiles: 50% of the population lie above the median, 50% below. We can also divide the population into *quartiles*: 25% lie below the first quartile, 50% below the second (which is also the median), 75% below the third quartile. There are 75 results in the previous table, so the first quartile will be the 19th result. Looking at the graph, this gives the first quartile as 11 and the third quartile (the 57th result) as 20. Similarly, the distribution can be divided into 100 parts knows as *percentiles*. "Your test result is in the top 5 percentiles of the population" means that at least 95% of people scored worse than you did.

There are various methods for calculating quartiles of a discrete distribution. For the exam, you will be expected either to use a graph or your GDC

Mode: The mode, or modal value, is simply the value which occurs the most often in a frequency distribution. In other words, the value with the greatest frequency. In a grouped frequency distribution, the group with the greatest frequency is called the *modal class*. Note that there can be more than one modal value or class.

Puzzle: Can you find five numbers such that mode < median < mean? And can you find five numbers such that mode < mean < median?

Solution: There are lots of possibilities, such as 2, 2, 5, 10, 12 and 2, 2, 6, 7, 8.

4.3 Measures of spread

The mean gives an indication of the "centre" of the distribution. The next most important statistic is a measure of "spread." For example, a buyer in a crisp factory testing different packing machines would be interested to know the mean number of crisps each machine put into bags, but it is equally important to know how *consistent* the machines are.

Standard deviation: The *standard deviation* provides a measure of how much results deviate, on average, from the mean.

Make sure you understand how to enter a frequency table into your calculator and how to obtain results for the mean and standard deviation.

Try calculating the SD of weight of peanuts in these 80 packets:

To save my typing fingers, I shall use SD as an abbreviation for standard deviation.

Weight	No of packets
$80 \leq W < 85$	5
$85 \leq W < 90$	10
$90 \leq W < 95$	15
$95 \leq W < 100$	26
$100 \leq W < 105$	13
$105 \leq W < 110$	7
$110 \leq W < 115$	4

You should find that the mean weight is 96.8 and the standard deviation is 7.41.

As a rough indicator, the majority of results in a reasonably symmetrical distribution are within two standard deviations of the mean (ie $\bar{x} \pm 2\text{SD}$). For example, a class takes a mathematics test. The mean score is 65% and the standard deviation is 8%. This means that most scores will be in the range 65 ± 16, ie 49% to 81%.

The *variance* is a useful statistic for further calculations, but does not have much significance on its own. It is the square of the standard deviation.

Outliers: The "two standard deviation test" gives us a useful way of identifying outliers.

Example: The following times, in seconds, were recorded in a race:

140, 148, 152, 155, 156, 156, 157, 160, 162, 162, 165, 170.

What evidence is there to suggest that 140 s was an exceptionally fast time for this group?

Solution: Using a GDC, mean = 156.9 and SD = 7.60, so 2SD below the mean is 141.7

140 s is therefore an outlier and can be classed as "exceptionally fast."

In the previous example this doesn't mean that the result can be discounted – some genuine results will be outliers. However, if the result had been recorded as 14 s this is clearly an error and should not be included in the data set.

Effect of changes to the data: Take a group of 10 children whose mean age is 12.4 years. What will be their mean age in 5 years' time? Since each of their ages will have had 5 added on, the mean will have increased by the same amount and will therefore be 17.4 years. And how will the standard deviation have changed? Not at all, since the *spread* of their ages around the mean will be exactly the same.

However, suppose a group of people take an exam marked out of 50; the mean is 35.2, and the standard deviation is 6.1. The scores are turned into percentages by doubling: the mean will now be 70.4, and the SD will have doubled as well to 12.2 since the marks will all have doubled their distance from the mean.

Thus, if a set of data has mean m and SD s, then the following rules apply:

These rules can be combined. If a set of data is doubled, and then 5 added on, the mean will be $2m + 5$ and the SD will be $2s$.

- Add a to each of the data values: the mean will be $m + a$, and the SD will be s.
- Multiply each of the data values by b: the mean will be mb, and the SD will be sb.

Not to be confused with the range which is simply the difference between the maximum and minimum values.

Interquartile range: The standard deviation of a distribution gives us a measure of the spread of the results which is calculated using each of the values. A cruder measure of the spread is the *interquartile range* which is calculated by subtracting the lower quartile from the upper quartile. Effectively, it tells us the spread of results for the middle 50% of the population. In questions, you will normally find the standard deviation "paired" with the mean, and the IQR paired with the median.

A survey is carried out to find the waiting time of 100 customers in a post office and the results tabulated as shown.

(a) Calculate an estimate of the mean waiting time.

(b) Construct a cumulative frequency table for the data.

(c) Use the table in (b) to draw a cumulative frequency graph, using a scale of 1 cm per 20 seconds on the horizontal axis and 1 cm per 10 customers on the vertical axis.

(d) Use the cumulative frequency graph to find estimates for the median and interquartile range.

Use the graph to estimate how many people waited more than 115 seconds.

Waiting time (sec)	Number of customers
0–20	5
20–40	18
40–60	30
60–80	22
80–100	9
100–120	7
120–140	6
140–160	3

(a)   Mean = 64.4 (GDC)

(b)

Waiting time (sec)	Cumulative frequency
≤ 20	5
≤ 40	23
≤ 60	53
≤ 80	75
≤ 100	84
≤ 120	91
≤ 140	97
≤ 160	100

(c)

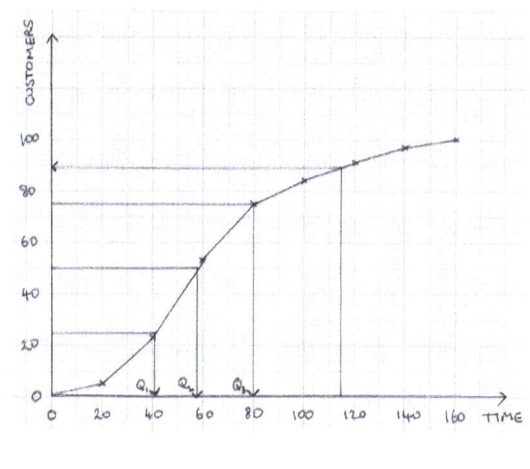

(d)   Median (also known as Q_2) = 58, Lower quartile (Q_1) = 41, Upper quartile (Q_3) = 80, IQR = 80 – 41 = 39

(e)   100 – 89 = 11 people waited more than 115 minutes

You must draw the relevant lines on the graph to show how you arrive at your answers. You will be given a little leeway with the numbers, but try to be as accurate as possible.

Outliers (again):   Since the IQR is a measure of spread we can also use it to define outliers. A well-used definition of an outlier is a data value which is more than $1.5 \times$ IQR above the upper quartile, or below the lower quartile.

In the question above, $Q_1 - 1.5 \times$ IQR = $41 - 1.5 \times 39 = -17.5$, so it is impossible for there to be outliers at the lower end of the distribution. But $Q_3 + 1.5 \times$ IQR = $80 + 1.5 \times 39 = 138.5$: any values above 138.5 can be considered as outliers. We know that there must be at least 3 outliers since there are 3 values above 140 minutes. *(This formula is not in your formula book.)*

Example: The cumulative frequency diagram shows the marks out of 70 gained by 80 students in a test. Write down the median, the quartiles and the IQR. One student took the test late and gained 66 marks – is this an outlier?

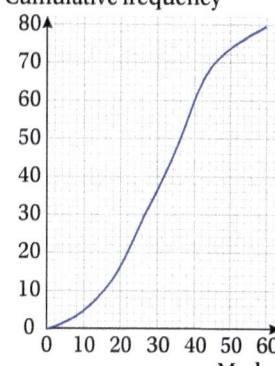

Cumulative frequency

Solution: Median = 32, Q_1 = 22, Q_3 = 40, IQR = 40 – 22 = 18.

Outliers would be greater than $40 + 27 = 67$. Therefore, the mark is not an outlier.

Box and Whisker plots: A box and whisker plot is a useful device for illustrating some key statistics for a distribution. The ends of the box represent the lower and upper quartiles, and the ends of the "whiskers" the extreme values. The median is shown by a line inside the box. A scale is drawn below the box and whisker plot,

An outlier would be indicated by a cross.

and different distributions can be compared. The illustration below shows the box and whisker plots for two math exams taken by a group of students.

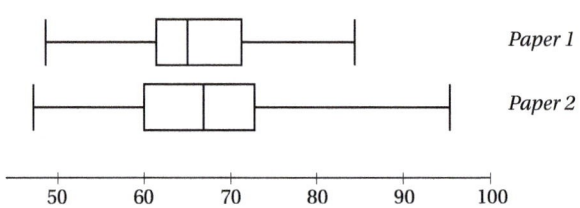

A question may ask you to compare the distributions – simple statements will suffice. For example:

- The range of results on Paper 1 is smaller.
- The median mark of the two papers is about the same.
- The interquartile range on Paper 2 is larger, so the results are more spread out.

Averages and Spread: Practice Exercise

Answers

1. (a) Discrete
(b)
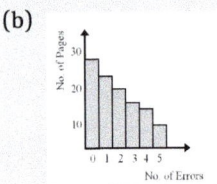

(c) 1.78   (d) 1.5   (e) 0

2. (a) 25   (b) 14
(c) $a = 20, b = 30$ (d) 25.6

3. (a) 6   (b) 6.1   (c) 1.22

4. (a) 48   (b) $a = 43, b = 80$
(c) 20

1.   The following table shows the number of errors per page in a 100 page document.

Number of errors	0	1	2	3	4	5
Number of pages	28	22	18	14	12	6

(a)   State whether the data is discrete or continuous.

(b)   Draw a bar chart to represent the data.

(c)   Find the mean number of errors per page.

(d)   Find the median number of errors per page.

(e)   Write down the modal number of errors per page.

2.   The cumulative frequency graph has been drawn from a frequency table showing the time it takes two hundred students to complete a computer game.

(a)   Find the median.

(b)   Find the interquartile range.

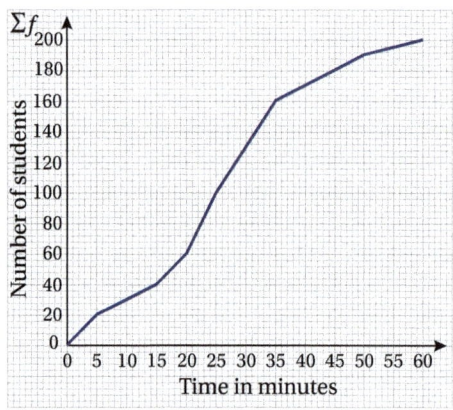

The graph has been drawn using the data in the following frequency table:

Time (min)	$0 < x \leq 5$	$5 < x \leq 15$	$15 < x \leq 20$	$20 < x \leq 25$	$25 < x \leq 35$	$35 < x \leq 50$	$50 < x \leq 60$
No. of students	20	20	a	40	60	b	10

(c)   Using the graph, find the values of a and b.

(d)   Calculate an estimate of the mean time taken to complete the computer game.

3. The bar chart shows the number of hours a professional musician practises each day during April:

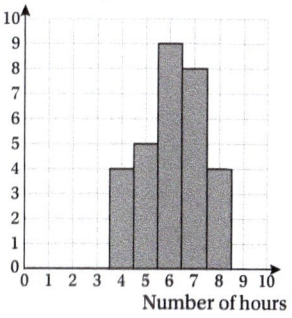

Number of days

   (a) Write down the modal number of hours.

   (b) Calculate the mean number of hours he practises each day.

   (c) Find the standard deviation.

Number of hours

4. The following diagram is a box and whisker plot (not to scale) for a set of data. The interquartile range is 15 and the range is 50.

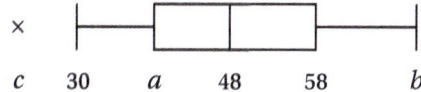

c 30 a 48   58 b

   (a) Write down the median value.

   (b) Find the values of a and b.

   (c) The cross represents an outlier. What is the highest possible integer value for c?

4.4 Sampling Methods

Ideally, statistics are gathered from a whole population. In practice, this is usually too expensive and too time-consuming, so a population sample is used instead. The difficulty is to make the sample representative of the population, so that conclusions drawn from the sample can be applied to the population.

"Population" doesn't just refer to people. For example, a machine in a factory may turn out 1000 components in a day. From this population, the quality control manager may want to select a sample of 20 components to test – how should he set about it?

One of the aims of sampling is to introduce as little bias as possible.

- Standing on a street corner at 11am selecting people for a survey introduces bias because the majority of the sample will not include people who are employed.

- Asking people to review a restaurant online with the promise of a possible discount introduces bias because people are more likely to give a good review.

- Selecting all 20 components (for the quality control test) in the morning may introduce bias because the machine could become more erratic later in the day.

Simple random sampling: The definition of a simple random sample is that every member of the population is equally likely to be chosen for the sample. Suppose you want a sample of 25 employees to be chosen from a company employing 300. Write all 300 names on pieces of paper, put them in a box, and then pull out 25 names. Every employee is thus equally likely to be in the sample. What this method will not achieve is a sample which reflects the make-up of the population: the male/female split, the number of employees in each age group and so on. By chance, the sample might contain all women aged between 25 and 30.

Instead of pieces of paper, every employee could be allocated a number, and then the sample is selected using a computerised random number generator.

Convenience sampling: In this method of sampling, data is collected from population members who are conveniently available. In other words, no inclusion criteria are specified before sampling takes place. Its main purpose is to gain some initial data prior

127

to a proper study taking place – for example, to obtain a perception of an image brand by simply going up to people in the street and asking their opinion.

Systematic sampling: Returning to our factory producing 1000 components a day, a simple way of selecting a sample of 20 is to choose every 50th component off the production line. If the first sample is chosen randomly from the first 50 components, then this ensures every component has an equal chance of being chosen. The main disadvantage of this method is that we need to know the population size to begin with. If, for example, a researcher wants to study a sample of 20 trees in a forest, she cannot systematically choose them without knowing how many trees there are in total.

Quota sampling: Quota sampling is used to select members of a population which has been divided into sub-groups. For example, a researcher wants to gather data on males and females, further sub-divided into age under 21 or 21 and over. He requires 30 of each (ie a sample size of 120), and simply approaches people until he has fulfilled each quota. This method is used if time or funding is limited but, since the sample is not genuinely random, the data is unreliable. Further bias could be introduced because the researcher may only question people who look approachable.

Stratified sampling: This method is the same as quota sampling except that within each group the sample is chosen by a method such as simple random sampling. For example, a school has 460 boys and 540 girls. A representative sample of 50 is to be chosen using stratified sampling. 50 is $\frac{1}{20}$ of the whole school, so we need to choose $\frac{1}{20} \times 460 = 23$ boys and $\frac{1}{20} \times 540 = 27$ girls. Now put all the boys' names in a box and choose 23 of them, then do the same for the 27 girls. Alternatively, go down the school list choosing every 20th boy and every 20th girl.

Stratified sampling will be used, for example, when it is important to an opinion pollster to gather results according to gender, age, political persuasion, and employment type.

The top two year groups in a school (Year 12 and Year 13) are to take part in a survey about the future of school uniform. The 200 pupils are divided by gender and year group as follows:

	Male	Female
Year 12	55	50
Year 13	48	47

A sample group of 30 are to be chosen. How would you select the group using stratified sampling?

There are 200 pupils in the group. $30 \div 200 = 0.15$.

    Year 12 male = $0.15 \times 55 = 8.25$ Select 8.
    Year 12 female = $0.15 \times 50 = 7.5$ Select 8.
    Year 13 male = $0.15 \times 48 = 7.2$ Select 7.
    Year 13 female = $0.15 \times 47 = 7.05$ Select 7.

Now allocate numbers to each of the members of the group, for example Year 12 males from 01 to 55, and use a random number generator to select the appropriate number from each group.

If the numbers don't come out as integers, it is important to ensure that the sample size from each group still adds up to the overall sample size. In this case I rounded 7.5 up to 8 to ensure the total was 30.

Any method of random sampling can be used for the second part.

4.5 Surveys and Questionnaires

Having organised a sample to test, thought must be given to the questions that are to be asked. A *survey* with simple answers, or a *multiple-choice* test, are both easy to analyse, but they do have disadvantages, such as:

- answers are restricted, and may not cover all the possibilities;
- if the question is not worded very carefully, it can be interpreted in different ways.

And, as with any survey, people may give answers they think the analyst would like to hear, rather a more truthful response.

On the other hand, more subjective questions are harder to analyse, particularly those referring to lifestyle. For example, "do you find your work stressful?" – would need much more amplification, since:

- it may change from day to day;
- it may depend on comparison with other people at work;
- somebody may find some aspects stressful, and not others.

Clearly, a straight "yes" or "no" would be pretty meaningless.

Selecting relevant variables: A survey may be assessing six different variables, let's say: Income, home-ownership, number of holidays per year, job satisfaction, number of children, favourite style of music. We want to look for correlation between any two pairs of these, and there are 15 possible pairs, so we do a χ^2 test for independence on each possible pair. At the 5% level, there is in fact a probability of more than a half that one of the pairs will show dependence by chance alone! Clearly, any positive result from one of the χ^2 tests needs further investigation.

See page 155 for the section on χ^2 tests.

Validity and Reliability: Whatever type of test is used to gather data, its designers need to consider how to make it valid and reliable.

A test is *valid* if it is actually measuring what it is supposed to be measuring, rather than something similar. As a trivial example, suppose we want to test people's knowledge of world capital cities, then a series of multiple-choice question about capital cities is clearly valid. It is said to have *content validity*. *Criterion validity* assesses whether a test reflects a certain set of abilities. This can be either *concurrent validity*, where the results are measured against a benchmark test, and high correlation indicates the test has strong validity; or *predictive validity*, where the results are tested against some future criterion. For example, the results of a set of general maths tests given to 50 students in Year 11 is then compared to their IB Maths Diploma grades attained in Year 13. If one of the tests correlates well, then that test could be used for future predictions (although if I were organising this I wouldn't be satisfied with just one set of results).

A test is *reliable* if it consistently produces similar results on each occasion it is used. There are two ways of testing reliability:

- *Test-retest:* Give the same test to the same group after a reasonable period of time has elapsed. It is reliable if there is strong correlation between the two sets of results.
- *Parallel forms:* Two tests are designed with similar sets of questions which aim to measure the same qualities. Both tests are given, after a short time has elapsed, to the same group of people. If the tests are reliable the results should correlate strongly.

Both of these methods have disadvantages. For test-restest, circumstances may have changed in the time interval between the tests. For parallel forms it is difficult to set up two tests which are exactly equivalent.

4.6 Probability Notation and Formulae

Notation: The *sample space* in a given situation is the set of all the things that can happen and is defined by the letter U. An *event* is one of the things that can happen and is given any other capital letter. A capital P stands for "probability", so we can shorten "the probability of event A" to P(A). The number of ways A can happen is denoted by $n(A)$. Probabilities are always numbers between 0 (definitely won't happen) and 1 (definitely will happen). The probability that A happens is given by $P(A) = \dfrac{n(A)}{n(U)}$. The probability that event A does **not** happen is denoted by A'. It follows that $P(A) + P(A') = 1$.

The set notation symbols $\cap$ and $\cup$ are used for the words "and" and "or" in probability.

Combined events: The probability of event A *or* event B happening (and this includes both) is calculated using addition.

- $P(A \cup B) = P(A) + P(B)$

but this formula works **only** if A and B are *mutually exclusive* – ie they cannot both happen. If they are not mutually exclusive, use:

- $P(A \cup B) = P(A) + P(B) - P(A \cap B)$

The probability of events A and B **both** happening is calculated by multiplication (remember that multiplying fractions gives a **smaller** answer and it is **less** likely that both events will happen than just one).

- $P(A \cap B) = P(A) \times P(B)$

A bag contains balls of two different colours. One is taken out, then another. The colour of the second is independent of the first if the first has been put back. If the first has been kept out, the colour of the second *depends* on the colour of the first.

but this formula works **only** if A and B are *independent* – ie one of them happening does not affect the probability of the other happening. If the events are not independent we are into the realms of *conditional probability* – ie the probability of one event happening if another has already happened. This is written as P($A|B$), and read as "the probability of A given B."

- $P(A|B) = \dfrac{P(A \cap B)}{P(B)}$

For the events A and B, $P(A) = 0.3$, $P(B) = 0.4$.

    (a)   Find $P(A \cup B)$ if A and B are independent events.

    (b)   Find $P(A' \cap B')$ if A and B are mutually exclusive events.

(a) $P(A \cap B) = 0.3 \times 0.4 = 0.12$
 	
    $P(A \cup B) = P(A) + P(B) - P(A \cap B)$
 	
             $= 0.3 + 0.4 - 0.12$
 	
             $= 0.58$
 	
(b) $P(A \cup B) = P(A) + P(B) = 0.7$
 	
    So, $P(A' \cap B') = 1 - P(A \cup B) = 0.3$
	(a) We are not told the events are mutually exclusive so we must use the full formula or P(A or B). This involves P(A and B) which we can calculate because they are independent.
	
	(b) is a new question so we cannot use independence. You will see in the next section how a Venn diagram can help you solve these sorts of problems more easily.

The formulae can be quite difficult to use, so only use them if you *have* to. Many probability questions can be solved by using appropriate diagrams as shown on the next few pages.

4.7 Lists and Tables of Outcomes

Lists: A list of possible outcomes is useful if there aren't too many of them; and it is important to ensure that each outcome in the list is equally likely. For example, when three coins are thrown, the possible combinations of heads and tails are:

HHH, HHT, HTH, HTT, THH, THT, TTH, TTT

If we want to find P(exactly two heads) we can see that there are three ways of achieving this (HHT, HTH, THH) so the probability is 3/8.

Possibility Space diagram: This is a way of showing a list of outcomes on a diagram, but can only be used for two events.

For example, the diagram below shows all the possible totals when two six-sided dice (red and green) are thrown:

Green

6	7	8	9	10	11	12
5	6	7	8	9	10	11
4	5	6	7	8	9	10
3	4	5	6	7	8	9
2	3	4	5	6	7	8
1	2	3	4	5	6	7
	1	**2**	**3**	**4**	**5**	**6**

Red

Thus there are 36 possibilities. Some examples of probabilities are:

P(Total of 5) = 4/36

P(Total of 5 or 7) = 10/36

P(Total of 4 or a double) = 8/36

P(Double | total ≥ 9) = 2/10

Note that there is only one way a 'double 2', say, can happen – a 2 on the green and a 2 on the red. But a 1 and a 3 can happen in two ways: 1 on the green and 3 on the red, or the other way around.

The conditional probability in the last example is easy to see on the diagram. We *know* that the total is ≥ 9, and there are 10 ways this can have happened. Of these, 2 could be a double.

Tables of outcomes: Tables of outcomes show how many ways two events can, or cannot, happen. For example, let's take a survey of 200 people of whom 90 are female. 60 people were unemployed, including 20 males. Filling that information into a table of outcomes, we get:

	Males	Females	Totals
Unemployed	20		60
Employed			
Totals		90	200

You will see that there is just enough information to allow us to fill in the rest of the table. Try it before looking at the answer below.

	Males	Females	Totals
Unemployed	20	40	60
Employed	90	50	140
Totals	110	90	200

Now, if a person is selected at random from the 200, what is the probability that the person is (a) an unemployed female, (b) a male, given that the person is employed.

(a)   There are 40 unemployed females out of 200, so P(unemployed female) $= \frac{40}{200}$.

(b)   Knowing that the person is employed, he/she must be one of the 140. Of these 90 are males, so P(male|employed) $= \frac{90}{140}$. As with the possibility space diagram, it is easy to deal with conditional probability when using a table of outcomes.

4.8 Venn Diagrams

In a room there are 20 people. 11 have black hair, 6 have glasses. 2 people have both black hair and glasses. Imagine that we draw two circles on the floor labelled "black hair" and "glasses" and ask the people to stand in the appropriate circle. The circles will have to overlap to allow for the two people with both. The numbers of people in each region of the room will be:

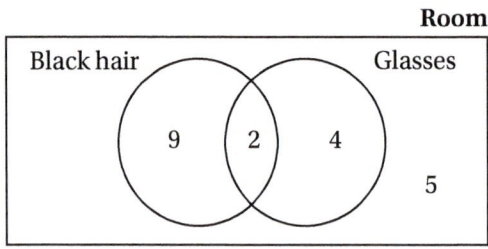

This is the same as a Venn Diagram. The "room" represents the sample space – for a particular question, there is nothing outside. Each circle represents a set, the overlap is the intersection.

Some examples of Venn Diagrams are shown below:

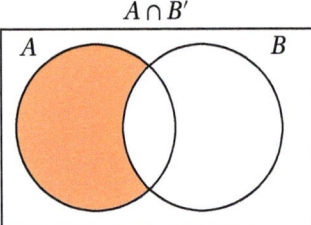
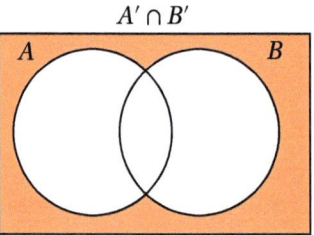
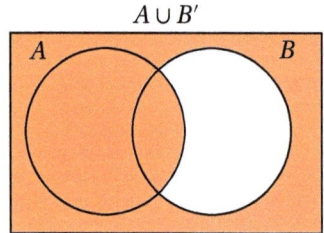

Points to note when filling in the numbers in a Venn Diagram:

- Start at the centre. If you are not told how many in the intersection, work it out like this: suppose you know there are 15 people in total in the two circles, 10 in circle A and 8 in circle B. 10 + 8 = 18, 3 more than 15, so there are 3 in the intersection.
- When we were told that there were 11 people with black hair, this *includes* those with both black hair and glasses. Same with the 6 people with glasses.

- Don't forget to fill in the outer region – although in some questions this set will be "empty."

Probabilities can now be calculated easily. When someone is selected at random, the probability they have:

Black hair and glasses = 2/20

Black hair and no glasses = 9/20

Not got glasses = 14/20

Glasses or black hair (or both) = 15/20

Glasses given black hair = 2/11

Glasses given not black hair = 4/9

Venn Diagrams are very helpful when calculating conditional probability – you may like to look at an article I have written for a fuller explanation.

See www.peakib.com

The next example shows how a Venn diagram can be used as an alternative to using the formulae.

Example: A and B are independent events. $P(A \cap B) = 0.2$, $P(A \cap B') = 0.3$

Find $P(A \cup B)$.

Solution: Intersections are easy to draw on a Venn diagram – see right.

Now we note that A and B are independent so we can use the formula

$P(A \cap B) = P(A) \times P(B)$.
$0.2 = 0.5 \times P(B) \Rightarrow P(B) = 0.4$

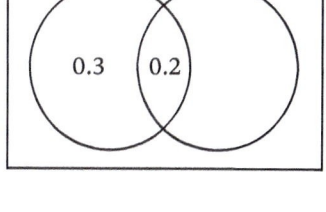

Using this information we can complete the Venn diagram, and hence answer the question:
$P(A \cup B) = 0.7$.

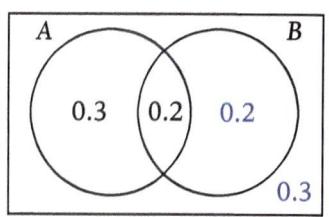

We can answer most questions once we have a complete Venn diagram. For example, can you show that $P(A|B) = 0.5$ and also that $P(B'|A) = 0.6$?

4.9 Tree Diagrams

Tree diagrams are used to work out the probabilities for a *succession* of events. To find the probability of a set of successive branches, multiply each individual probability *along* the branches. To find the probability of one of several branches occurring, add the probabilities of each outcome.

Note that the probabilities associated with, say, taking two balls out of a bag simultaneously are the same as if the balls were taken out consecutively.

eg: P(rains today) = 0.3.
If it rains today, P(rains tomorrow) = 0.65

However, if it is dry today, P(rains tomorrow) = 0.2
The tree diagram shows the full set of possible outcomes and their associated probabilities.

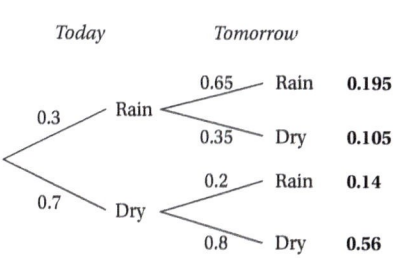

Note the following points:

- Probabilities of branches coming out of one point add to give 1 since they cover all possibilities.
- The overall probabilities also add to give 1.
- The weather tomorrow is *not* independent of the weather today, hence the different probabilities depending on today's weather.

Some example probabilities are:

- P(two rainy days) = 0.195
- P(at least one rainy day) = 0.195 + 0.105 + 0.14 = 0.44

$$= 1 - P(\text{two dry days})$$

- P(exactly one rainy day) = 0.105 + 0.14 = 0.245

Questions about tree diagrams often come with a sting in the tail in the form of a conditional probability problem.

Consider the following tree diagram where event A = "my alarm clock works" and event L = "I am late for school."

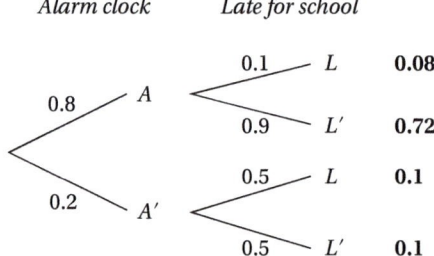

P(I am late for school) = $P(L|A) + P(L|A') = 0.08 + 0.1 = 0.18$.

But suppose I am late for school, and my teacher says: "Obviously your alarm clock didn't work", what is the probability she is right?

The calculation is $P(A'|L) = \dfrac{P(A' \cap L)}{P(L)} = \dfrac{0.1}{0.18} = 0.556$. Another way to think of this is to consider the expected values over 100 days. On 18 of them we would expect to be late, and of these 18 we would expect the alarm clock to have failed on 10.

The probability that it is sunny in Bonn is 0.6. The probability that a girl passes a test is 0.7 when it is sunny and 0.3 when it is not sunny.

(a) Create a tree diagram to represent the above information.

(b) What is the probability that the girl passes the test on any given day?

(c) Given that a girl passes the test, what is the probability that it is sunny in Bonn?

(b) 0.54 (c) 0.778

See worked solution online

Expected number of occurrences: In the previous example about alarm clocks, I suggested looking at the number of times I would expect to be late over a period of 100 days. P(late) = 0.18, so I would expect to be late on $0.18 \times 100 = 18$ days. This is a specific example of a simple calculation: whenever you have the probability p of an event, and that event could happen on n occasions, the expected number of occurrences is pn.

4.10 Discrete Probability Distributions

A probability distribution shows the probabilities for all the outcomes of a particular event. Discrete probability distributions relate to events which can only have certain outcomes – usually in the form of integers.

Uniform distributions: If all the outcomes are equally likely, the distribution is called uniform. For example, here is the probability distribution for the random variable X where X represents the outcomes when throwing a die.

x	1	2	3	4	5	6
P($X = x$)	1/6	1/6	1/6	1/6	1/6	1/6

Note that the capital letter X is used to describe the random variable, whereas lower case x is used to represent the actual values.

Distributions defined by a function: The following is an example of a probability distribution defined by a function:

$$P(X = x) = \begin{cases} kx, \ x = 1, 2, 3, 4, 5 \\ 0 \text{ otherwise} \end{cases}$$

This means that x can only take values 1 to 5, and has probability kx for these values. The best thing to do is put all the information into a table:

x	1	2	3	4	5
P($X = x$)	k	$2k$	$3k$	$4k$	$5k$

In all probability distributions, the probabilities add to give 1, so $15k = 1$, giving $k = \frac{1}{15}$. We can fill the probabilities into the table:

x	1	2	3	4	5
P($X = x$)	$\frac{1}{15}$	$\frac{2}{15}$	$\frac{3}{15}$	$\frac{4}{15}$	$\frac{5}{15}$

Expected value (mean): By multiplying each value of x by its associated probability, we obtain the *expected mean*. Thus the formula is: $E(X) = \sum xp$. In the above example we get $\frac{55}{15} = 3.67$, and the more times we carry out the trial, the closer the **actual** mean will get to this value.

Example: The probability distribution for a random variable X is given by:

$P(X = x) = kx(x - 1)$, for $x = 2, 3, 4, 5, 6$

(a) Find the value of k; (b) Find the expected mean of the distribution.

Solution: First we must draw up the probability table:

x	2	3	4	5	6
P($X = x$)	$2k$	$6k$	$12k$	$20k$	$30k$

Thus $70k = 1 \ \Rightarrow \ k = \frac{1}{70}$.

Now we can fill the probabilities into the table and work out the expected mean.

x	2	3	4	5	6
$P(X = x)$	$\frac{2}{70}$	$\frac{6}{70}$	$\frac{12}{70}$	$\frac{20}{70}$	$\frac{30}{70}$

$$E(X) = 2 \times \frac{2}{70} + 3 \times \frac{6}{70} + 4 \times \frac{12}{70} + 5 \times \frac{20}{70} + 6 \times \frac{30}{70} = \frac{350}{70} = 5$$

The following table shows the probability distribution of a discrete random variable X:

x	−1	0	1	2	3
$P(X = x)$	0.2	$10k^2$	0	0.4	$3k$

(a)   Find the value of k

(b)   Find the expected value of X.

(a) $10k^2 + 3k + 0.6 = 1 \Rightarrow k = 0.1$ (GDC)

(b) $E(X) = -0.2 + 0 + 0.8 + 0.9 = 1.5$

Games of chance: Let's play a game. You throw two dice. If you get a 9 or 11, I'll give you $3; if you get a double, I'll give you $2. The catch is, you must pay me $1 to play. Is it worth it? We can draw up a table of probabilities (see page 131 for how to deal with totals of two dice).

Event	Prob.	Outcome
9 or 11	6/36	$3
Double	6/36	$2
Other	24/36	$0

The expected mean is $\frac{6}{36} \times 3 + \frac{6}{36} \times 2 + \frac{24}{36} \times 0 = \frac{30}{36}$. Thus, on average, you can expect to win under $1 per game, so you will lose out in the long run – and you will decline my offer to play. (Moral: you can't make money out of IB Mathematics students!) You could alternatively include the $1 in the table by making the outcomes $2, $1 and –$1. This would make the expected mean $-\frac{6}{36}$.

4.11 Expectation Algebra

Linear transformation of a single random variable: Consider the set of values 1, 2, 3, 4, 5. The mean is 3 and the standard deviation is 1.414. If we add 4 to each value, the new mean will be 7, but the standard deviation will remain unchanged (because the *spread* of results around the mean is exactly the same). However, if we multiply the results by 2, the mean will multiply by 2 and so will the standard deviation; the spread is two times greater than it was.

The algebra normally deals with variance, but the results are easier to understand in terms of standard deviation.

I have demonstrated the results with some actual data, but we shall apply them to a random variable – that is, with *expected* mean and variance.

Suppose we multiply by 2 then add 4. In summary:

$X = \{1, 2, 3, 4, 5\}$ $2X + 4 = \{6, 8, 10, 12, 14\}$

$E(X) = 3$ $E(2X + 4) = 2E(X) + 4 = 10$

$Var(X) = 1.414^2 = 2$ $Var(2X + 4) = 2^2 Var(X) = 8$

In general, $E(aX + b) = aE(X) + b$ and $Var(aX + b) = a^2 Var(X)$. Just remember that adding a constant to a random variable has no effect on the variance.

Linear combinations of several random variables: When independent random variables are combined, we can calculate the combined expectation and variance:

$$E(a_1X_1 \pm a_2X_2 \pm ...) = a_1E(X_1) \pm a_2E(X_2) \pm ...$$

$$Var(a_1X_1 \pm a_2X_2 \pm ...) = a_1^2 Var(X_1) + a_2^2 Var(X_2) + ...$$

> Note that the variances are added even when the variables are subtracted

Be careful to distinguish between the random variable $2X$ and the random variable $X_1 + X_2$. For example, when a single die is thrown, let X be the face value. $E(X) = 3.5$, $Var(X) = 2.917$. If we double the scores, then $E(2X) = 2E(X) = 7$ and $Var(2X) = 4Var(X) = 11.67$. (You might like to enter the values of X and $2X$ as lists on your calculator, and use the Stats functions to confirm these values). However, if you throw the die a second time and add the scores together, each throw is independent: the total score will be a new random variable, $X_1 + X_2$. $E(X_1 + X_2) = E(X_1) + E(X_2) = 7$, as before, but $Var(X_1 + X_2) = Var(X_1) + Var(X_2) = 5.832$.

An important point to remember is that combinations of Normal distributions are themselves Normal.

Example: An end of year exam consists of a written paper marked out of 80, and an aural test marked out of 10. To calculate the overall result the aural mark (A) is multiplied by 2 and added to the written result (W). The marks for both the tests are normally distributed. $W \sim N(68, 7.2)$ and $A \sim N(8.1, 2.2)$ for the written and aural tests respectively. What is the distribution of the overall mark?

Solution: In this case, the aural mark has been multiplied by 2, so the distribution is $T = W + 2A$. The mean is calculated as $68 + 2 \times 8.1 = 84.2$, and the variance will be $7.2 + 4 \times 2.2 = 16$. Thus, $T \sim N(84.2, 16)$

Example: Apples are packed into boxes of 24. The weights (in g) of the apples are normally distributed with $X \sim N(120, 20)$, and the weight of the box also forms a normal distribution, $B \sim N(200, 30)$. What is the distribution of the total weight, T, of the box of apples?

Solution: $T = (X_1 + X_2 + ... + X_{24} + B)$.

Thus, $E(T) = 24 \times 120 + 200 = 3080$ g, and $Var(T) = 24 \times 20 + 30 = 510$ g.

$\Rightarrow T \sim N(3080, 510)$.

Quick practice: Expectation Algebra

1. A random variable Y has a mean of 4 and a variance of 1.5. Calculate the expected value and variance of $2Y$, $3Y + 1$, $Y - 3$.

2. Independent random variables S and T are such that $E(S) = 12$, $Var(S) = 3$, $E(T) = 5$ and $Var(T) = 1.4$. Find the expected value and variance of $(S + 2T)$ and of $(2S - 3T)$.

3. From past experience, the temperatures recorded at a weather station have an expected mean of 22.4°C and an expected variance of 2.3°C. Given that the conversion from Centigrade to Fahrenheit is $F = \frac{9}{5}C + 32$ find the expected mean and variance measured in Fahrenheit.

4. A clothes shop has pairs of socks priced at £2.00, £3.00 and £3.50, and shirts priced at £12.00, £14.00 and £15.00. Assuming all the items sell equally well, find the mean price and the variance for socks and for the shirts. If I buy 4 pairs of socks and 2 shirts, find the expectation and variance of the amount I pay (assuming I am equally likely to choose each item).

Answers

1. 8, 6; 13, 13.5; 1, 1.5.

2. 22, 8.6; 9, 24.6.

3. 72.3, 7.45.

4. Socks: 2.83, 0.329; Shirts: 13.67, 1.556; Combined: 38.66, 12.448

4.12 The Binomial Distribution

The binomial probability distribution is a special case of a discrete distribution. You can use it when:

- there are a fixed number of "trials";
- each trial has only two possible outcomes, "success" and "failure";
- the results of each trial are independent of each other;
- the probability of success remains the same.

For example, my young child wakes me up 1 night in 4. I want to find the probability that I will be woken up 3 nights out of 10.

We must assume that the probability of being woken on any one night is independent of being woken on any other night.

- The number of trials, n, is 10.
- The probability of "success" (ie being woken up!) is 0.25
- We therefore say that the distribution is $X \sim B(10, 0.25)$

Using the binomial distribution calculator on the GDC, we find that P($X = 3$) = 0.250. Note that probabilities are normally given to 3SF.

Also note that "all or nothing" cases can be calculated as simple powers: P(woken up all 10 nights) = 0.25^{10}; and the probability of not being woken up at all in ten nights is 0.75^{10}.

Getting the probability of success: You may simply be given the probability of success, or:

- you calculate the probability from previous experience (as in the example above);
- you calculate it from your knowledge of the situation (eg: success is getting a 2 on the spinner shown in the margin: P = 1/3);
- the probability is the result of a calculation from a previous part of the question.

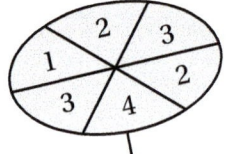

This question assumes that a prizewinner can win more than one prize. Why?

Example: 350 of the 500 pupils in a school have the letter "s" in their name. Six sports prizes are awarded at the end of term. What is the probability four of the prizewinners have an "s" in their name.

Solution: The first sentence gives us the probability of success.

P("s") = $\frac{350}{500}$ = 0.7. Therefore $X \sim B(6, 0.7)$.

P(4 successes) = 0.324 (GDC)

More than one outcome: Since binomial probabilities are all mutually exclusive (I cannot be woken up both 3 nights *and* 4 nights in 10), the probability of one of several outcomes occurring can be found by addition. Thus, P(I am woken up 3 or 4 nights out of 10) is:

$$P(X = 3) + P(X = 4) = 0.250 + 0.146 = 0.396$$

Check the wording of questions carefully. It might say: "Find the probability that I have at least eight nights when I am *not* woken up." Check this also gives 0.526.

Cumulative probabilities: What is the probability of being woken up on fewer than 3 nights out of 10: that is, P(0, 1, or 2). You can add these three probabilities together or use the cumulative probability function on your calculator which gives 0.526. This enables us to answer questions such as: "Find the probability that I am awoken on at least 3 nights out of 10."

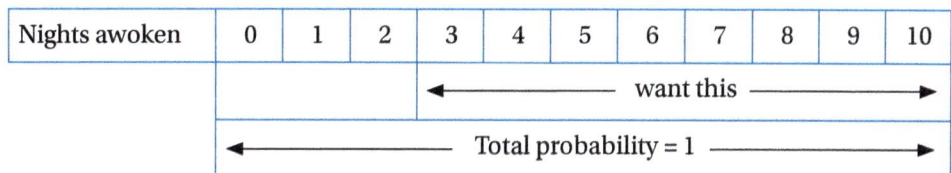

Nights awoken	0	1	2	3	4	5	6	7	8	9	10
				← ——————— want this ——————— →							
	← ——————————— Total probability = 1 ——————————— →										

The diagram shows that the easiest way to calculate this is to find the cumulative probability up to 2, and subtract the answer from 1. This gives $1 - 0.526 = 0.474$.

On some GDCs this cumulative probability could be calculated directly.

Joe is a football player. When shooting penalties, he succeeds 3 times out of every 5. In practice, he shoots 8 times. Find, to 3 significant figures, the probabilities of:

  (a)   Scoring all 8 penalties.

  (b)   Scoring 6 penalties out of 8.

  (c)   Scoring at least 6 penalties out of 8.

$X \sim B(8, 0.6)$

  (a) $P(X = 8) = 0.6^8 = 0.0168$

  (b) $P(X = 6) = 0.209$ (GDC)

  (c) $P(X \geq 6) = 1 - P(X \leq 5)$

        $= 0.315$ (GDC)

It's always worth stating the distribution you are going to use – the examiner can see what you're doing, and it helps sort things out in your mind too.

In part (c) I have used the distribution functions on the GDC, but I have also shown some working to ensure method marks in case my answer is wrong.

I've included the next question here although it could well be part of a paper 2 question because it illustrates how to deal with wording which can send your mind in a spin. The key thing is to strip the wording down until it becomes possible to see the binomial probabilities required.

A machine contains a critical component. This component is replicated 10 times within the machine, and the machine works as long as at least one of the ten components is working. Each has an independent probability of failing within one year of 0.7, and all the components are replaced at the end of a year.

(a) Find the probability that all 10 fail within the year.

(b) Find the probability that the machine is in operation at the end of the year.

(c) (i) Suppose we put in n components. What is the probability that the machine is operating at the end of the year?

(ii) Hence find the smallest number of components to install which will ensure a probability of at least 0.99 that the machine is working at the end of the year.

(a) $X \sim B(10, 0.7)$ where X = component fails $P(X = 10) = 0.7^{10} = 0.0282$ (b) P(machine works) = $1 - P(X = 10) = 0.9718$ (c) (i) $X \sim B(n, 0.7)$ P(machine works) = $1 - 0.7n$ (ii) $1 - 0.7n > 0.99$ $n > 12.9$ So the smallest value of n is 13	*(a) The only issue is a semantic one – the failure of a component is a probability success!* *(b) At least 1 component working is the same as 1 – P(none are working); a familiar calculation in binomial probability questions.* *(c) Restating the distribution parameters helps to see the connection between part (c) and part (b).*

It is known that 1 out of 20 printed circuit boards supplied by a certain manufacturer has a fault. What is the probability that at least 1 in a batch of 10 is faulty?

Why might a binomial probability not be appropriate? Refer to the conditions on page 138 under which a binomial distribution is valid. Firstly, the events must be **independent**. The answer to the question in the notes box is 0.401. However, the assumption of independence may be wrong: the 1 in 20 is an average figure over a period of time, but perhaps if the temperature in the factory rises too much, more faulty boards are produced. Then our batch of 10, if they were all manufactured together, may have a higher incidence of faults.

Secondly, you cannot use the binomial distribution if the probabilities change. For example, there are 10 pieces of paper folded up in a box, and three have crosses marked on them. To find the probability that, when two pieces of paper are drawn out, neither has a cross, you need a tree diagram. The probabilities change each time a piece of paper is removed.

In reality, as he tires, his probability of success would probably decrease.

Expected mean: Fortunately, we do not have to go through the normal process for discrete distributions – there is a simple formula for the expected mean of a binomial distribution. Suppose Joe (who appeared a few questions back) decides to enter a marathon penalty shooting competition and goes for 400 shots. How many times would he succeed? His probability of success is 0.6, so we would expect him to succeed $0.6 \times 400 = 240$ times. Thus, if $X \sim B(n, p)$, then $E(X) = np$.

Remember that variance is the square of standard deviation.

Expected variance: It turns out that, not only is there a simple formula for the expected mean of a binomial distribution, there is also an equally simple one for the expected variance.

If $X \sim B(n, p)$, then $\text{Var}(X) = np(1 - p)$. Returning once again to Joe, the expected variance of his 400 penalty shots will be $400 \times 0.6 \times 0.4 = 96$. Thus the expected standard deviation will be $\sqrt{96} = 9.80$. What does this tell us? You should recall that we generally expect results to be within two standard deviations of the mean. In this case, with a mean of 240, this gives a likely range of about 221 to 259. If Joe scored, say, 270 penalties out of 400, we might need to question the accuracy of the 3 out of 5 figure quoted in the original question. He could be better than we thought!

In the next example we see again how a binomial distribution question leads to a bit of algebra.

Example: A binomial $X \sim (n, p)$ distribution has mean 4 and variance 2.4. Find the values of n and p.

Solution: When you have to find two unknowns, the chances are you will end up with simultaneous equations. These will arise from the formulae for the mean and the variance.

$$np = 4,$$

$$np(1 - p) = 2.4$$

Substitute the value of np from the first equation into the second:

$$4(1 - p) = 2.4 \Rightarrow p = 0.4$$

Substitute this value of p into the first equation:

$$n = \frac{4}{0.4} = 10$$

Binomial Distribution: Practice Exercise

1. $X \sim B(7, 0.25)$. Find $P(X = 5)$, the mean and variance of X.

2. $X \sim B(12, 0.8)$. Find $P(X = 10)$, $P(X > 8)$ and $P(X$ is less than 7$)$.

3. Which is more likely: (a) $X \sim B(10, 0.15)$, $P(X = 3)$
   (b) $X \sim B(12, 0.12)$, $P(X = 3)$?

4. A coin is thrown 50 times. X represents the number of heads. What would be a likely range of values for X?

5. I work 5 days a week, and I'm late home from work about once every ten days. My partner gets cross if I'm late home more than once in a week, and I then buy them a present. How many weeks in a year would I expect to have to buy a present?

Answers

1. 0.0115, 1.75, 1.3125
2. 0.283, 0.795, 0.0194
3. 0.13, 0.12, so (a)
4. 17.9 to 32.1, so 18 to 32
5. $P(X > 1) = 0.081$. In a year, this happens $52 \times 0.081 = 4.2$. So about 4 times a year.

4.13 The Poisson Distribution

Conditions for a Poisson distribution: A random variable has a Poisson distribution if the following conditions are fulfilled:

- The variable is discrete
- The occurrences are random
- The occurrences are independent
- There is a known mean rate for the occurrences

Examples of events which might be modelled by a Poisson distribution are: the number of telephone calls a receptionist receives in a 5 minute period; the number of deaths per year from lightning strikes.

Calculating Poisson probabilities: One of the main features of the Poisson distribution is that the mean is equal to the variance. The symbol used for both is λ, so

$$E(X) = Var(X) = \lambda$$

To calculate the probability that $X = r$:

$$P(X = r) = \frac{\lambda^r e^{-\lambda}}{r!}, \text{ for } \lambda > 0 \text{ and } r = 0, 1, \dots$$

You will not need the Poisson formula – but I included it because I'm rather fond of it!

Some questions may require use of the formula, but otherwise use your GDC to calculate Poisson probabilities.

Cumulative probabilities: The receptionist above receives a mean of 4.4 calls in a 5 minute period. If we want to find the probability of 2 or fewer calls in 5 minutes, then use the Poisson cumulative probability function to get the answer 0.185.

Unlike binomial probabilities, there isn't a set number of trials. Theoretically, there is no limit to the number of calls in a 5 minute period.

If we have been asked to find the probability of **more** than 2 calls in a 5 minute period, this can be calculated as $1 - 0.185 = 0.815$.

Weak spots occur at random in the manufacture of a certain type of rope at an average rate of 2 per 100 metres.

(a)   If X represents the number of weak spots in 100 metres of cable, write down the distribution of X.

(b)   Lengths of rope are wound onto drums. Each drum carries 60 m of rope. Find the probability that the drum will have 4 or more weak spots.

(c)   A contractor buys 5 such drums. Find the probability that exactly two have at least one weak spot each.

(a) $X \sim P_0(2)$	*(a) In exam questions you may be expected to identify which is the most suitable distribution to use.*
(b) $\lambda = 0.6 \times 2 = 1.2$	
$P(X \geq 4) = 1 - P(X < 4) = 1 - 0.96623\dots = 0.0338$	*(b) Be **very** careful with the inequalities. I have used Poisson cumulative up to X = 3.*
(c) $P(X \geq 1) = 0.6988$	
If Y is the number of drums with at least one weak spot, $Y \sim B(5, 0.6988)$	*(c) 2 out of 5 must be a binomial probability, but first we must use the Poisson distribution to find the probability of "success".*
$P(Y = 2) = 0.133$ to 3SF	

Sum of two Poisson distributions: If two Poisson distributions, with means λ and μ, are added together then the combined distribution is also Poisson and has mean $\lambda + \mu$. For example, two switchboard operators receive calls both modelled by a Poisson distribution with means of 3 calls every 10 minutes and 4.5 calls every 10 minutes. The overall number of calls X received can also be modelled by a Poisson distribution where $X \sim P_0(7.5)$.

4.14 The Normal Distribution

The Normal Distribution is used to model many commonly occurring frequency distributions, eg: the heights of trees, weights of people. The curve has the following properties:

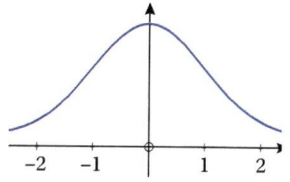

- It is symmetrical about the mean value, μ.

- The median is the same as the mean.

- The curve approaches the x-axis asymptotically (although this is not true for the majority of distributions the curve is modelling).

The curve (shown in the diagram) is called the *standard* normal distribution: its mean is 0, its standard deviation is 1 and the area under the curve is 1.

Mean and standard deviation: The basis of all normal distribution calculations is how many standard deviations above or below the mean a particular value is. Thus, if a group of people have a mean height $170\,\text{cm}$ with standard deviation $10\,\text{cm}$, and a mean weight of $65\,\text{kg}$ with standard deviation $5\,\text{kg}$, then the probability that a person chosen at random is less than $180\,\text{cm}$ high is exactly the same as the probability of weighing less than $70\,\text{kg}$; both values are one standard deviation above the mean.

In general, approximately two thirds (nearer 68%) of the data are within 1 standard deviation of the mean. This can be written as $\mu \pm \sigma$. 95% of the data are in the range $\mu \pm 2\sigma$ and 99.7% in the range $\mu \pm 3\sigma$. Using the data above, we could say that the majority of the group of people have weights between $65 \pm 10\,\text{kg}$; or from $55\,\text{kg}$ to $75\,\text{kg}$. And it would be very unlikely to find anyone with a weight less than 50kg or more than $80\,\text{kg}$.

Normal probability calculations: Normal probabilities can be calculated on your GDC by entering four values: lower bound, upper bound, mean, standard deviation. In questions where there is no lower or upper bound, you can either use values such as -1×10^{99} and 1×10^{99}, or just any values much smaller or larger than those in the question.

For example, if $\mu = 35$ and $\sigma = 3$, and we need to find $P(X > 40)$, then an upper bound of 100 is easily big enough.

When finding answers on your GDC you need to show some working. I suggest a shaded diagram will demonstrate to the examiner that you understand what is happening.

Example: A group of people are asked to carry out a simple task. The length of time taken, in minutes, follows a normal distribution where the mean is 3.2 and standard deviation is 0.6.

If a member of the group is chosen at random, find:

This would be written as:
$X \sim N(3.2, 0.6^2)$

    (a) P(she takes between 2.5 and 3 minutes)

    (b) P(she takes more than 3 minutes)

Solution: (a) On your GDC, enter the values for the lower and upper bounds, the mean and the SD.

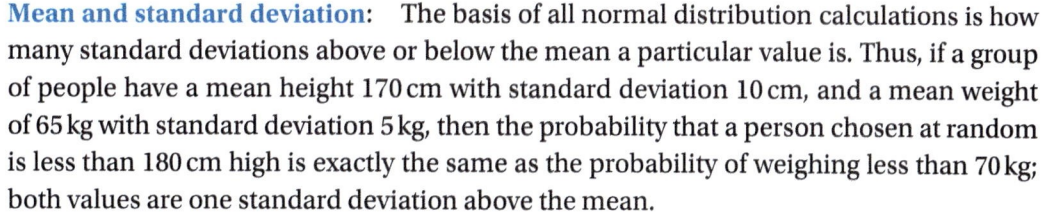

$P(2.5 \leq X \leq 3) = 0.248$

    (b) Now enter 3 for the lower bound, and a number such as 100 for the upper bound.

$P(X > 3) = 0.631$

You can also use the inverse normal function on the GDC to reverse the process. The inverse will give you the X value associated with a particular probability.

Carrying on with the previous question:

The fastest 10% of the group are then selected to perform a more advanced task.

(c) What was the slowest time required to be selected?

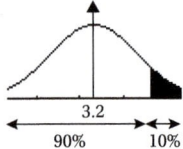

Solution: The inverse function can be used only from the left hand end of the distribution. Thus, to find the lowest value for the top 10%, we must find the highest value of the bottom 90%. Entering this value into the GDC, along with the mean and SD, we find the slowest time of the fastest 10% is 3.97 minutes.

The next example illustrates why you must be careful using normal distribution calculations when the required answer is an integer.

Example: In a certain exam, 12% of candidates were ungraded. If the mean mark was 52 and the standard deviation was 13, and the marks were normally distributed what is the highest mark which a candidate could obtain and not gain a grade, assuming marks are integers?

Solution: Using the inverse normal function on the GDC, we find that the bottom 12% of candidates scored up to 36.7 marks. Now, 36.7 rounds to 37 – but a candidate scoring 37 would have been graded. So the highest possible mark an ungraded candidate scored is 36.

A particular component which is required on an assembly line is manufactured such that its mean weight is 85 g with standard deviation 0.4 g.

(a) Explain why you would not expect a component to be manufactured weighing 86.5 g.

(b) Find the probability that a component weighs less than 84.7 g.

(c) 100 components are chosen at random. How many would be expected to weigh more than 84.7 g.

(d) The lightest 10% and the heaviest 10% of components are rejected. Calculate the range of weights of components which are accepted.

(a) $\mu + 3\sigma = 86.2$ g

86.5 g is therefore over 3 SDs above the mean.

(b) $P(X < 84.7) = 0.227$ (GDC)

(c) $P(X > 84.7) = 1 - 0.227 = 0.773$

$100 \times 0.773 = 77.3$

Therefore expected number which weigh more than 84.7 g is 77.

(d) Top 10% weigh more than 85.5 g (GDC)

By symmetry, bottom 10% weigh less than 84.5 g. Therefore components in the range

$83.5\,g < X < 85.5\,g$ are accepted.

(b) A quick sketch helps you see what is going on, and shows the examiner what you are doing.

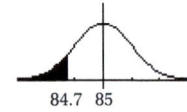

(c) Symmetry is often a powerful tool when answering normal probability questions.

Normal Distribution: Practice Exercise

1. For a normal distribution with mean 25 and standard deviation 3.5, find (a) $P(X < 26)$, (b) $P(24 < X < 27)$

2. The mean of a normal distribution is 1.7, and we are given that $P(X < 1.85) = 0.74$. Find $P(1.55 < X < 1.85)$.

3. The weight of 200 apples in a sack is normally distributed with mean 120 g and standard deviation 8 g. Find the probability that an apple weighs more than 110 g; and also the number of apples in the sack that weigh less than 100 g.

4. A set of exam results X is distributed normally with mean 65 and standard deviation 12. Where should the grade boundary be set such that no more than the top 15% of students gain an A grade in the exam?

Answers

1. (a) 0.612 (b) 0.329

2. 0.48

3. 0.895; 21.

4. 78

Links to other probability techniques: Probability questions do not necessarily fall into neat categories – here is a tree diagram question, there is a binomial distribution question, and that one is a normal probability question. Quite often, a question begins by asking you to calculate a probability from, say, a normal distribution, but then you might need to use that probability in a binomial distribution; you do need to recognise just what you are being asked.

Example: 40 players regularly train with the Griffins basketball squad. Their heights are normally distributed with mean 193 cm and standard deviation 4.8 cm.

     (a) Find the probability that a member of the squad is taller than 196 cm.

     (b) A team of 5 is chosen for a match. What is the probability that at least 3 of them are taller than 196 cm?

Solution: (a) Using a straightforward normal distribution calculation on the GDC, we find that $P(X > 196) = 0.266$.

     (b) At least 3 out of 5? This is a binomial probability question with $Y \sim B(5, 0.266)$. In other words, the normal probability we have just calculated is the "success" probability for the binomial.

         $P(Y \leq 3) = 1 - P(Y \geq 2) = 1 - 0.879 = 0.121$

4.15 Correlation

Scatter diagrams: Two sets of data which appear to have a relationship between them are said to be *correlated*. For example, a company may find that there is a direct relationship between the amount it spends on advertising and its sales figures. Note that correlation does not imply causality: the correlation may be coincidental, or it may be linked to a third factor (perhaps, in this case, differing economic conditions).

A simple way to assess possible correlation is to draw a *scatter diagram*. The two sets of data are plotted on a standard x-y graph (but not joined in any way).

Qualitative conclusions which can be drawn about the correlation are:

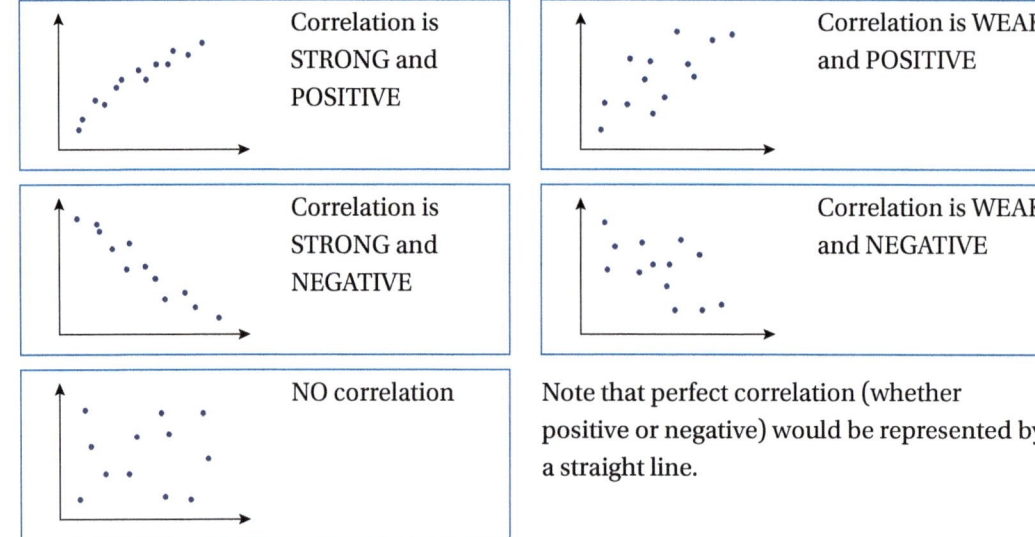

Correlation is STRONG and POSITIVE

Correlation is WEAK and POSITIVE

Correlation is STRONG and NEGATIVE

Correlation is WEAK and NEGATIVE

NO correlation

Note that perfect correlation (whether positive or negative) would be represented by a straight line.

Let's look at a couple of examples:

In this first one, a group of 10 to 16 year old boys were timed running 100m. We can see that there is strong negative correlation between their ages and their times.

Do you think that you can extrapolate to estimate the time taken by a 21 year old? A 60 year old?

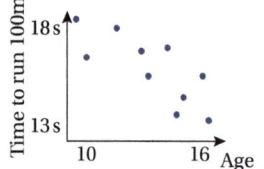

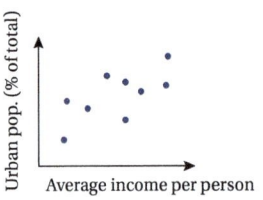

In this second graph, we are comparing countries by the percentage who live in cities against the average income per person in each country. Quite strong correlation – but, again, could we extrapolate and say that the figures will always increase together?

The answer in both cases is "no." In the first example, extrapolation would indicate that a 40 year old could run 100 m in 0 seconds! I would expect the graph to begin to turn soon, and then start back up again. In the second example, it's possible that extrapolation would work for a short distance (although we haven't been given scales); however, no country can have more than 100% of its people living in cities.

Line of best fit: A scatter diagram indicates the relationship between two variables. If we conclude that there *is* a relationship, we can draw in the "line of best fit" by eye and then use this to predict more pairs of values. If you know the mean values of the two variables, the line of best fit should pass through the point $(\bar{x}, \bar{y})$. Note that 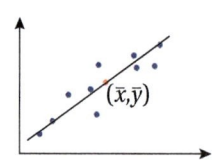 although *interpolation* (ie putting new points in between existing points) is fairly safe, *extrapolation* (ie continuing the line beyond the existing points) may not be valid. There may be reasons why the relationship does not continue in the same way.

Correlation Coefficient: For a quantitative assessment of correlation we can calculate the *Pearson product-moment coefficient*, denoted by r. This is derived from all the pairs of values and has the following properties:

- A coefficient of –1 indicates perfect negative correlation.
- A coefficient of 0 indicates no correlation.

- A coefficient of +1 indicates perfect positive correlation.

The size of r (ie the positive value of r) indicates the strength of the correlation, but this also depends on the number of pairs of values. However, we can say generally that:

- $0.25 \leq r < 0.5 \Rightarrow$ weak correlation
- $0.5 \leq r < 0.75 \Rightarrow$ moderate correlation
- $0.75 \leq r < 1 \Rightarrow$ strong correlation

(and similarly for negative values of r).

Using your calculator: Although you may have learnt how to calculate the equation of the line of best fit (or *regression line*) and also correlation coefficients using formulae, you will not be expected to do this in the exam. However, you should be able to do both of these things using your GDC. Generally, the method is to input the pairs of x and y values, then use the appropriate calculator functions.

Check that you are able to carry out these calculations using the following data:

x	2	4	5	7	9	10	11	15
y	3	4	6	6	7	9	10	11

You should find that the correlation coefficient is 0.97 (not surprising when you look at how closely the y values follow the x values) and that the regression line of y on x has equation $y = 1.89 + 0.65x$.

> Because the regression line is "y on x" it can only be used to calculate y values given x values.

The equation of the line can be used to predict further data points. For example, what is the likely value of y when $x = 7.8$, and when $x = 17$?

When $x = 7.8$, $y = 1.89 + 0.65 \times 7.8 = 6.96$

When $x = 17$, $y = 1.89 + 0.65 \times 17 = 12.94$

However, as mentioned earlier, the latter result must be treated with caution since it has been extrapolated beyond the end of the known data – there is no guarantee that the relationship between x and y will continue to hold.

Since the equation of the regression line is a linear function, the values of a and b can represent physical quantities. a is the gradient, so represents "y quantity per x quantity". And b is the y-intercept, so represents the value when $x = 0$, which is less likely to be meaningful. Using the first example above (time to run 100 m against age in years), a would represent "time to run 100 m per year"; in other words, how much faster a boy would run for every extra year in age. b would represent "time to run 100 m at age 0" which is clearly a nonsense; in any case, the data is only valid between ages 10 and 16.

The following table shows the amount of fuel (y litres) used by a car to travel certain distances (x km).

Distance (x km)	50	80	125	160	195
Fuel (y litres)	4.6	7.1	10.9	13.9	16.9

This data can be modelled by the regression line with equation $y = ax + b$.

(a) (i) Find the values of a and b.

(ii) Explain what the gradient a represents.

(b) Use the model to estimate the amount of fuel used to drive 100 km.

(c) Could the model be used to estimate the amount of fuel to drive 250 km?

a) (i) $a = 0.085$, $b = 0.326$

(ii) litres/km travelled

b) 8.82

c) Yes, because fuel usage doesn't change with distance.

OR

No, because this would involve extrapolating beyond the data range.

a) It's very easy to test whether your answer is correct – just try using the formula on one of the x values and check you get the corresponding y value (or something close to it).

c) Either answer would do – as long as you give a valid explanation. But it's perhaps safer to always say no to extrapolation.

Spearman's rank correlation coefficient: A rank correlation coefficient gives us another way of testing the correlation between two quantities. For example, you might want to test whether there is any correlation between the results of a history exam and a French exam, and this can be done by comparing each pupil's rank in the two exams.

As with Pearson, a rank coefficient of 1 means you have perfect correlation - ie each pupil ranked exactly the same in the two exams. A coefficient of –1 is perfect negative correlation, and a coefficient of 0 means no correlation at all.

The procedure to calculate the Spearman rank correlation coefficient on your GDC is *exactly* the same as for the product-moment coefficient, except that your two lists contain the ranks, not the actual values.

Let's continue with our exam example; here is the table of results for 10 pupils:

History	56	75	45	71	61	64	58	80	76	61
French	66	70	40	60	65	56	59	77	67	63

Now let's rank the pupils in each exam. Note that if two pupils have the same rank we use the average of the two ranks.

History	9	3	10	4	6.5	5	8	1	2	6.5
French	4	2	10	7	5	9	8	1	3	6

Enter these ranks into two lists on your GDC, and find the correlation coefficient. For Spearman, we use the symbol r_s. In this case $r_s = 0.669$ which, despite being a small sample, is quite strong correlation.

Comparison of Pearson and Spearman:

- Pearson is only testing for linearity. In other words, the coefficient tells you how close the relationship is to a straight line.

- Spearman tests the correlation between ranks irrespective of whether or not there is an underlying linear relationship. The diagram shows an example of a non-linear relationship between two quantities which will have a strong rank coefficient.

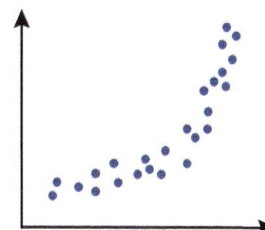

- Spearman is less sensitive to outliers since it only tests ranks. (compare the difference between mean and median).

- Both give correlation coefficients in the range –1 to 1.

Piecewise models: When a function is graphed as separate line segments, meeting at common points, it is known as a piecewise linear function (see figure on right). Sometimes correlation on a scatter diagram would be better achieved with more than one line of best fit, in which case we end up with a piecewise model. In other words, a single regression line may fit reasonably well, but two (or more) would fit the points much more closely.

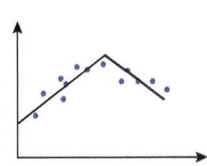

Here's an example of a piecewise model with two clearly defined sections. It's important that the point where the lines join satisfies the equations of both of the two regression lines. What sort of situations might give rise to a piecewise model? Well, the scatter diagram could represent measures of air pollution at different distances from a city centre. Moving out from the centre we pass through an increasingly industrialised zone with greater pollution; and then further out in the suburbs, so pollution decreases.

Example: A piecewise linear model contains two regression lines with equations $y = 1.6x + 2$ for $0 \leq x \leq 2$, $y = 7 - 0.9x$ for $2 \leq x \leq 5$. Find the point where the two lines meet, and also calculate y values when $x = 0.8$ and when $x = 3.5$.

> Or you can tell from the domains in the question that the two lines meet when $x = 2$.

Solution: Solving the two equations simultaneously (GDC) we find the lines meet at $(2, 5.2)$. When $x = 0.8$ (first section), $y = 4.28$. When $x = 3.5$ (second section), $y = 3.85$.

4.16 Non-linear Regression

The relationship between two quantities is not necessarily linear: the scattergraph example in the paragraph above comparing Pearson with Spearman shows points which seem to follow a curve. The question is, which curve? If we can make a guess at the underlying function (eg: exponential, cubic), we can then use the GDC to suggest a best fit curve, and also to provide a measure of how well the curve fits the observations using the *coefficient of determination*.

With your GDC to hand, work through an example with me which should cover all the techniques you need.

Example: In a chemical reaction, the mass of substance A decreases. The table shows the mass, in g, measured every minute over a 5 minute period.

Time (mins)	0	1	2	3	4	5
Mass (g)	54	40	29	24	20	18

(a) It is thought that the mass fits an exponential model with equation $m = ak^t$. Use your GDC to find the values of a and k.

First enter the values into two lists. Then use the appropriate stats function which finds the best exponential regression curve.

$a = 49.7$, $k = 0.801$, so $m = 49.7 \times 0.801^t$.

(b) Write down the coefficient of determination and explain its significance.

$r^2 = 0.967$. This means that 96.7% of the variation in the dependent variable is predictable from the model. This value shows that the model is a very good fit.

(c) Plot the points on a scattergraph and draw the exponential curve.

Although you can do this without the GDC, it may be useful to know how to plot points on a scattergraph. This is what I get.

The GDC also shows us which points are above and which are below the curve. First set the graph window using the data in the table. The plot feature is then used to plot the points, and the standard graph drawing functionality used to draw the graph.

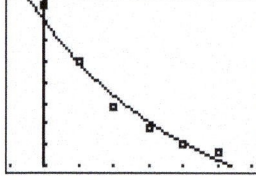

(d) By drawing a line on your graph, carry out a calculation to find when the mass of substance A has been reduced to half its initial amount.

We need to know when there are only 27g left, so we draw the line $y = 27$ on the graph and find the point of intersection. From the table, we can see that the answer must be between 2 and 3 minutes, and from the GDC we get: $t = 2.75$ mins.

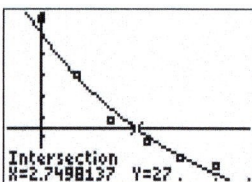

You may also be asked to find the *sum of the square residuals.* The residual for any of the points is the difference between the y-coordinate, and the y-value predicted by the model. Square the residual to ensure a positive number, and then add them for all the points to get a measure of how close the curve is to all the points. This can be done by simply carrying out the calculation for each point, or it can be done directly on the GDC as follows:

The exact process will depend on the make of GDC. These instructions give a general outline.

- Ensure that the coordinates of the points are entered as List 1 and List 2.
- Define List 3 as a function, the model (in the example above, the exponential function). Thus List 3 has the predicted y values.
- Define List 4 as $(\text{List } 3 - \text{List } 2)^2$.
- Use the GDC's list functionality to sum List 4.

In the example above, the sum of the square residuals (written as SS_{res}) is 32.05.

Other models you may be asked to work with are quadratic, cubic, power and sine.

The following table shows the rate of water flow, F litres/sec, compared to nozzle pressure, P psi in a water hose.

Pressure (psi)	11	17	20	25	40	55
Flow (litres/sec)	21.1	32.0	35.3	37.1	41.8	47.6

Hans suggests that there might be a linear relationship between rate of flow and water pressure.

    (d) (i)   Find the equation of the regression line in the form $F = a + bP$

        (ii) Calculate the sum of square residuals.

Betty proposes a better model using a power function.

    (e) (i)   Find the equation of a curve in the form $F = aP^b$ to fit the data.

        (ii) Calculate the sum of the square residuals.

    (f) State which model most closely fits the data, giving a reason for your answer.

(a) (i) $F = 21.7 + 0.503P$

     (ii) $SS_{res} = 64.3$

(b) (i) $F = 8.219P^{0.452}$

     (ii) $SS_{res} = 41.9$

(c) Betty's model is a closer fit because the lower value of SS_{res} indicates the points are closer to the curve.

The corresponding values of R² are 0.842 and 0.887, also indicating that the power function is a better model that the linear function.

In practice it would be a good idea to plot the points to see if the linear model is more likely.

4.17 Sample Means

Life can get a little bit confusing from here on: it's really important that you understand when to use each of the methods and formulae which follow so that you can confidently tackle questions. In particular, check whether you are dealing with a *population* or a *sample*.

Distribution of the sample mean: In most practical situations it is impossible to measure values for a complete population, and statisticians must therefore deal with samples. How are the mean and standard deviation of a sample related to those for the parent population? Clearly, if you take a number of samples then the means of these samples will themselves form a distribution: this is called the *sampling distribution of the means*. Using μ and σ^2 for the population mean and variance, and n for the sample size, it can be shown that the sampling distribution also has mean μ but has variance $\frac{\sigma^2}{n}$. The sampling distribution of the means is given the symbol $\bar{X}$, so we can write:

$$E(\bar{X}) = \mu, \ Var(\bar{X}) = \frac{\sigma^2}{n}$$

Note that the larger n is, the lower the variance. It follows that larger samples give better estimates of the population mean. Fairly obvious, perhaps, but this is the proof that larger samples give more accurate results.

Example: A sample of size 8 is taken from a population with distribution N(10,4). Find the probability that the sample mean is more than 11.

Solution: $\bar{X}$ will be normally distributed (because the parent population is), so

$$\text{Var}(\bar{X}) = \frac{\sigma^2}{n} = \frac{4}{8} = 0.5 \quad \text{Thus } \bar{X} \sim N(10, 0.5).$$

$$P(\bar{X} > 11) = 0.0786.$$

The standard deviation of the sample means is $\frac{\sigma}{\sqrt{n}}$ and this is known as *the standard error of the mean.*

The random variable X is normally distributed such that $X \sim N(40, 12)$. The mean of a random sample of n observations of X is denoted by $\bar{X}$. Find:

(a)   The distribution of $\bar{X}$ when $n = 10$.

(b) $P(\bar{X} \geq 41)$ when $n = 10$.

(c)   The smallest value of n for which $P(\bar{X} \geq 41) < 0.05$.

(a) The distribution of sample means is $N\left(\mu, \frac{\sigma^2}{n}\right)$ $= N(40, 1.2)$.	*The first two parts are straightforward bookwork.*
(b) Thus $P(\bar{X} \geq 41) = 0.181$ to 3SF (GDC)	*Part (c) is a typical "work backwards" question.*
(c) We must solve an equation on the GDC using 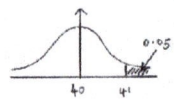 the normal distribution functionality. With lower bound –1000 (say), upper bound 41, mean 40 and SD $\sqrt{\frac{12}{x}}$ the probability is 0.95. This gives $x = 32.47$, so $n = 33$	*Since n has to be an integer, we then have to decide whether the answer is 32 or 33. A value of 32 would lead to a smaller value of the variance, and so a larger probability.*

Central limit theorem: If the parent population is normally distributed then, unsurprisingly, so is the distribution of the sample means. In other words, if you take a large number of samples, you would find that the means of these samples would be themselves normally distributed, centred on the population mean. However, if the parent population is **not** normal, the sampling distribution of the mean is **still** approximately normal, and becomes more so the larger the sample size n. This result is called the *Central Limit Theorem*.

Estimators: Usually a population is too large for all values to be measured; hence the mean and variance of a population are generally unknown. However, the mean and variance of a *sample* can be used to provide estimates. Generally, the sample mean is a good estimate of the population mean. But the *spread* of values in a population is going to be greater than in a sample, so the population standard deviation (and hence variance) is also going to be greater than the sample standard deviation. It can be shown that, if s_n is the sample standard deviation, then $\sqrt{\frac{n}{n-1}} \, s_n$ serves as a good estimate of the population standard deviation, where n is the sample size. The larger the value of n, the better the estimate.

In summary:

- To estimate μ use $\bar{x}$
- To estimate σ^2 use $\dfrac{n}{n-1}s_n^2$

Example: Five measurements of the volume of acid required in a titration are 25.1, 25.2, 25.2, 25.0 and 25.5 cm³. Use these results to obtain point estimates for the mean and standard deviation of volume of acid required.

Solution: These results form a sample from which we must estimate the mean and standard deviation of the population of all possible measurements. A "point" estimate means a single value, as opposed to a range (see confidence intervals). Check that the mean and s.d. of the five measurements are 25.2 and 0.0280....

Therefore, the required estimates are:

$$\mu = 25.2 \text{ cm}^3, \sigma = \sqrt{\tfrac{5}{4}} \times 0.0280 = 0.0350 \text{ cm}^3.$$

4.18 Confidence Intervals

Confidence intervals for the population mean with known σ: A sample mean is an estimator for the population mean, but if you take several samples, each will have a different mean. We need to find a range either side of a sample mean within which we can be fairly confident the population mean lies: the surer we want to be, the wider the range must be.

The sample means form a normal distribution with parameters $E(\bar{X}) = \mu$, $Var(\bar{X}) = \dfrac{\sigma^2}{n}$. 95% of the values lie within 1.96 standard deviations above and below the mean, so we can see that 95% of the sample means lie in the range $\mu - 1.96 \times \dfrac{\sigma}{\sqrt{n}} < \bar{X} < \mu + 1.96 \times \dfrac{\sigma}{\sqrt{n}}$.

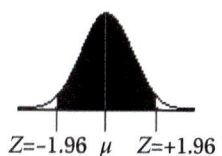
$Z=-1.96 \quad \mu \quad Z=+1.96$

Rearranging this gives $\bar{X} - 1.96 \times \dfrac{\sigma}{\sqrt{n}} < \mu < \bar{X} + 1.96 \times \dfrac{\sigma}{\sqrt{n}}$, and we can be 95% certain that the population mean lies in this range.

The range is called the *confidence interval*, and the upper and lower values are called *confidence limits*. This is best seen in the diagram which shows 5 sample means with their confidence intervals. Most of the time the population mean lies within the confidence interval, but in one case it lies outside. With a 95% confidence level, this would only happen 5 times out of 100. Other confidence levels can of course be used.

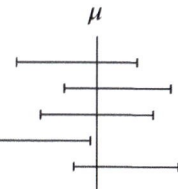

However, you will not be required to use the formula since your GDC will do the heavy lifting for you. You only need to enter the sample mean and SD, the sample size and the confidence level.

Example: A plant produces steel sheets whose weights are known to be normally distributed with a standard deviation of 2.4 kg. A random sample of 30 sheets had a mean weight of 31.4 kg. Find 99% confidence limits for the population mean.

Solution: $30.3 < \mu < 32.5$ (GDC)

The *t*-distribution is similar to the normal distribution

Confidence intervals for the population mean with unknown σ: In practice, it is likely that the population standard deviation is not known. In such cases, we use a procedure based on the *t*-distribution to calculate the confidence intervals. This will use the estimator for population standard deviation which we have seen is:

$$s_{n-1} = \sqrt{\frac{n}{n-1}}\, s_n$$

It therefore follows that we must use given data values for the sample, in which case the calculation on the GDC is carried out using "data" rather than "stats".

For example, suppose tests are being carried out on a new sleeping pill. Ten people are given the sleeping pill, and the number of hours sleep they get is:

7.5   8.1   7.7   6.9   9.2   8.5   8.1   7.6   7.6   6.8

Assuming the number of hours sleep is normally distributed, find a 95% confidence interval for the mean length of time someone would sleep after taking the sleeping pill.

Since we don't know the population standard deviation we must use the *t*-interval calculation, and we start by entering the data into a list. Try this on your GDC, and you should get $7.29 < \mu < 8.31$ (to 2DP).

Sample Means and Confidence Intervals: Practice Exercise

Answers

1. 0.0331
2. 28
3. 235.9, 8.98, 0.242
4. 0.284, 0.0453
5. [9.59, 10.01]
6. [2.18, 2.68]

1. A sample of size 6 is taken from a Normal distribution N(10, 2^2). What is the probability that the sample mean exceeds 11.5?

2. Samples of newly manufactured steel girders are taken to check their lengths. If the standard deviation of the lengths of all girders is 2.6 cm, find the smallest sample that must be taken for the standard error of the mean to be less than 0.5 cm.

3. The number of parts per million of pollutants in 7 randomly chosen samples of air in a city centre are:

    240.8   237.3   236.6   234.2   233.9   232.5

    Calculate estimates of the mean and variance of the population from which this sample was taken, given that the parent population is normally distributed, and hence find the probability that there are more than 238 parts per million.

4. Before testing, 10% of computer chips are known to be defective. Find the probability that fewer than 9.5% out of a sample of (a) 1000 and (b) 10000 are defective.

5. 50 measurements of the acceleration due to gravity, *g*, had a mean value of 9.8 ms^{-2} and a standard deviation of 0.75 ms^{-2}. What are the 95% confidence limits for *g*.

6. Families in the state of Ibonia generally have large numbers of children. The government carries out a survey of 100 families and asks each of them how many children there are in the family. The results were:

No. of children	0	1	2	3	4	5	6
No. of families	8	22	27	19	13	8	3

Calculate a 90% confidence interval for the mean number of children per family.

A computer manufacturing company buys large quantities of hard disks from several suppliers. Hard disk quality is checked by a process called RTT which gives results on a continuous scale from 0 to 100. Based on previous experience the company assumes that the results are normally distributed with a mean of 68 and a standard deviation of 3.

(a) Find the probability that a hard disk selected at random has a result less than 67.

Shipments arrive from suppliers on a daily basis. A sample of 10 hard disks is taken from each shipment at random and tested. If the mean of the sample is more than 67, the shipment is accepted, otherwise it is rejected.

(b) Find the probability that a shipment is rejected.

There is a \$1000 penalty each day that a shipment is rejected. A particular supplier's hard disks have a mean of 67.5 and a standard deviation of 2.8.

(c) (i) Find the probability that this supplier's shipment is accepted.

(ii) Calculate the expected penalty per 6-day week for this supplier to the nearest dollar, assuming there is a shipment each day.

(a) $X \sim N(68, 3^2)$

$P(X < 67) = 0.36944... \approx 0.369$

(b) Distribution of sample means $S \sim N\left(68, \frac{3^2}{10}\right)$

$P(S < 67) = 0.1459... \approx 0.146$

(c) (i) For this supplier, $S \sim N\left(67.5, \frac{2.8^2}{10}\right)$

$P(S > 67) = 0.7138... \approx 0.714$

(ii) The number of rejections R in a 6-day week is binomially distributed with $R \sim B(6, 0.286)$

$E(R) = 6 \times 0.286 = 1.7168....$

∴ Expected weekly penalty

$= 1.7168 \times 1000 = \$1717$

Parts (a) and (b) show clearly the difference between calculating a probability for a population and for a sample.

Be careful of the wording in (c): we are told the probability of acceptance, but need to find the probability of rejection.

In part (c) we are still dealing with samples, although in this case it's all the samples from one supplier. Part (ii), the sting in the tail, requires you to use a binomial distribution since we need the expected value of an event over 6 "trials". "Success" (in probability terms) is rejection, hence P(success) = 1 − 0.714.

4.19 The Chi-squared (χ^2) Test

Contingency Tables: A contingency table is a sort of Venn diagram which shows how a population splits according to two factors. We would like to see whether or not the two factors are independent of each other. The procedure is to take a table of observed frequencies and, from this, draw up a table of *expected* frequencies on the assumption of independence.

For example, suppose we take a group of 80 people split into 3 age groups, and we are interested to see how many of each have gained penalty points on their driving licence during the last year.

	Age < 21	21 ≤ Age ≤ 60	Age > 60	Total
No points	24	19	21	64
Points	10	1	5	16
Total	34	20	26	80

Working out the expected frequencies: We can see that one-fifth of the group gained points: we would therefore expect, for example, 4 of the middle age group to have gained points. 4 is the expected frequency, and the calculation to work it out is $\frac{16 \times 20}{80}$; in general, the calculation is $\frac{\text{row total} \times \text{column total}}{\text{grand total}}$. (Note that the numbers do not necessarily come out as integers.) The table of expected frequencies is calculated as:

The newly calculated figures are shown in blue

	Age < 21	21 ≤ Age ≤ 60	Age > 60	Total
No points	**27**	**16**	**21**	64
Points	**7**	**4**	**5**	16
Total	34	20	26	80

So the number of under-21s with penalty points is more than we expected – could it be this high by chance alone, or could it be that the two factors (age and points) are in fact related?

χ^2 **test**:   We now need a statistic which measures by how much the observed frequencies differ from the expected frequencies. This is called the χ^2 statistic; the higher it is, the further the observed values deviate from the expected values, and so the less likely that chance alone is involved. In the exam you will need to be able to enter the observed frequencies into your GDC, and then use the stats functions to calculate the statistic. In this case, $\chi^2 = 4.7$

Find out how to enter the observed frequencies into your GDC. It can then calculate the expected frequencies and the value of χ^2.

On its own, this doesn't tell us much. We now need a *critical value* for χ^2 which tells us at which point "could have happened by chance" becomes "unlikely to have happened by chance." This critical value depends on the number of *degrees of freedom* and also on the *significance level* we set for the test.

Degrees of freedom: The more rows and columns we have, the more pairs of values there will be. So the value of χ^2 will be higher, and so will the critical value. The number of "degrees of freedom" is calculated as (rows – 1) × (columns – 1); in our example, this is $(2 – 1) \times (3 – 1) = 2$. The term degrees of freedom does not need to be understood, but you must be able to calculate it!

Significance level: How certain do we want to be that the value of χ^2 could not happen by chance alone? A significance level of 5% means that only in 5% of cases would you get a value this high or higher by chance alone. This in turn means that if our value of χ^2 is above the critical value, we can be at least 95% certain that the things we are measuring are in fact related. If we want to be at least 99% certain, we go for a significance level of 1%.

...but at the 10% level, a less stringent test, the result is significant, and we could say that the number of points is related to age.

The test: In the exam you will be given the critical value if required. In our case, for 2 degrees of freedom, the critical value is 4.605 at the 10% level, 5.991 at the 5% level and 7.378 at the 1% level. So, if we had set the level at 5%, $\chi^2 = 4.7$ is ***not*** significant, and the observed frequencies could have happened by chance. There is no evidence to suggest that age and penalty points are related.

4.20 Hypothesis Tests – Contingency Tables

A hypothesis test begins by setting up the *null hypothesis* (which is what we believe to be true at the moment, the status quo) and an *alternative hypothesis* (which is the opposite of the null hypothesis). Data is collected and analysed, the analysis being carried out according to what exactly we want to test. We then calculate from the sample how unlikely the null hypothesis is compared to a pre-stated significance level, and make our conclusion accordingly.

Let's see how the results from the contingency table in the previous section can be bundled up into a hypothesis test. The null hypothesis, H_0 is that penalty points are independent of age, so the alternative hypothesis, H_1, must be that points are not independent of age.

> H_0: Penalty points are independent of age
>
> H_1: Penalty points are not independent of age
>
> Significance level = 5%
>
> $\chi^2 = 4.7$
>
> Degrees of freedom = 2
>
> Critical value = 5.991
>
> $4.7 < 5.991 \Rightarrow$ accept H_0
>
> Conclusion: There is no evidence to suggest that the number of penalty points depends on age.

There is often confusion on how we interpret the χ^2 statistic compared to the critical value. Remember that χ^2 is a measure of differences between observed and expected values, so if it is less than the critical value, we interpret that by saying that the differences are not so great that they couldn't have happened by chance alone.

---

200 people from different areas in a country were asked which was their preferred radio station. The results are shown in the table below:

	Radio X	Radio Y	Radio Z	TOTALS
North	42	13	35	90
Central	30	10	20	60
South	13	17	20	50
TOTALS	85	40	75	200

A χ^2 test for independence is to be carried out at the 5% level.

  (a)   Write down the null and alternative hypotheses.

  (b)   What is the expected number of Radio Y listeners in the Central area?

  (c)   Find the χ^2 value.

  (d)   Given a critical value of 9.488, state whether you would accept or reject the null hypothesis.

  (e)   In terms of the 5% level, what conclusion do you draw?

(a) H_0: Preferred radio station is independent of where people live

H_1: Preferred radio station is not independent of where people live

(b) $\frac{40}{200} \times 60 = 12$

(c) 11.39

(d) $11.39 > 9.488$, therefore reject H_0

(e) The evidence suggests that people living in different areas prefer different radio stations. The probability of the observed frequencies happening by chance alone is less than 5%.

Part (b) could also be found by looking at the table of expected frequencies on your GDC. However, if the question had asked you to calculate the number of listeners, then the answer would have to be as shown.

If you are using p-values you do not need to know the value of χ^2

p-value: For a particular value of χ^2, and with a particular number of degrees of freedom, the p-value measures the probability that the test statistic is at least as high as the one actually obtained. For the question above, the p-value is 0.0225 (obtained from the GDC). Put simply, the lower the p-value, the more extreme the statistic, and therefore the more likely that the quantities being measured are not independent. Since the p-value is less than 5%, this also suggests support for H_1.

Contingency table hypothesis test: Practice question

A certain type of electronic component is made in three different factories. Samples were taken from each to see how many were defective.

Fill in the totals in the contingency table below which shows the results of the sampling:

	Satisfactory	Defective	Totals
Factory A	97	3	
Factory B	65	5	
Factory C	65	15	
Totals			

(a) What is the probability that a component chosen at random:

(i) Came from Factory B?

(ii) Was a defective from Factory C?

(iii) Was satisfactory given that it came from Factory A?

(b) Complete the table below with all the expected frequencies, giving answers to the nearest whole number.

	Satisfactory	Defective	Totals
Factory A		9	
Factory B			
Factory C	73		
Totals	227		250

A χ^2 test is carried out to see if the proportion of defective components is different at each factory.

(c)   Write down the null and alternative hypotheses.

(d)   Write down the χ^2 statistic.

(e)   Write down the number of degrees of freedom.

(f)   At the 1% level, the critical value is 9.21. What conclusions can be drawn?

Answers:

Row totals: 100, 70, 80; Column totals: 227, 23; Grand total: 250.

(a)   (i) 0.28;   (ii) 0.06;   (iii) 0.97.

(b)   A/Satis: 91; B/Satis: 64; B/Defect: 6; C/Defect: 7.

(c) H_0: The number of defectives is independent of the factory.

   H_1: The number of defectives is not independent of the factory.

(d)   13.7

(e)   2

(f)   13.7 > 9.21 so reject H_0. There is evidence to suggest that some factories produce more defectives than others.

> Because of rounding errors, you may find that the totals don't add exactly to give the grand total.

> Note that the p-value is 0.001 which is less than 1%

4.21 Testing for the Mean of a Normal Distribution

Suppose that a teacher believes that Year 8 pupils are taller than they were 10 years ago. To test this, she finds out the mean height 10 years ago and she then takes a sample of *this* year's Year 8 pupils. A hypothesis test can then be constructed to see how likely it is that this sample could still have come from a population of Year 8 pupils with the same mean as before.

Other typical examples:

- Patients have a 60% chance of recovering from a particular illness. A new drug is tested on 100 patients and 68 recover. Could this have happened by chance? If we find it is very unlikely then we can accept the fact that the drug is having a positive effect.

- A cereal manufacturer claims that his packets contain a mean weight of cereal of 200 g. A rival samples 50 packets and finds the mean weight is only 190 g. If this is significantly less than the nominal mean, then this will cast doubt on the manufacturer's claim.

In all the tests that follow, if the population standard deviation is known we use the z-test (normal distribution), otherwise we use the t-test. Also note that for small samples we must assume they come from a normally distributed population; with large samples it doesn't matter.

***Known* σ:**   The following example takes you through the process:

The IQ scores of a population are normally distributed with a mean of 100 and a standard deviation of 15. A psychologist wishes to test the theory that eating chocolate improves your score. A random sample of 60 people is selected and they are each given a bar of chocolate before sitting an IQ test. Their mean score is 103. Test the psychologist's theory at the 5% level.

Stage 1: Set up the hypotheses.

The *null* hypothesis is the status quo; in this case, that the mean IQ of the 60 people is 100. The *alternative* hypothesis is that the chocolate is having an effect, and that the mean score has increased. Write the hypotheses like this:

H_0: $\mu = 100$

H_1: $\mu > 100$

Stage 2: Enter the stats onto your GDC.

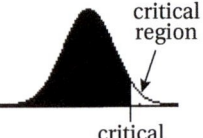

critical region

critical value

The diagram in the margin shows the "unlikely" region (ie the top 5%) of a normal curve. The value where this occurs is called the *critical value*, and the region is called the *critical region*.

Enter the population mean and sd, the sample mean and sample size. Also enter the test type: $\mu > \mu_0$. The resulting *test statistic* is $z = 1.549$ (ie 103 is 1.549 standard deviations above 100), and the *p*-value is 0.0607.

Stage 3: State the conclusion.

The *p*-value tells us that a sample mean IQ this high could happen in 6.07% of samples. Since 6.07% > 5%, we accept H_0 and, perhaps reluctantly, conclude that there is no evidence that chocolate increases IQ.

Set up a 2-tailed test on your GDC by selecting $\mu \neq \mu_0$.

One-tailed or two-tailed?   In the example above, the claim was that chocolate increased IQ, so we were only testing if the mean was higher than normal. This was a *one-tailed test*. However, if the claim was that chocolate affected IQ (either way), then H_1 would have been $\mu \neq 100$, and the test would be *two-tailed*. In practice, the only difference is that the 5% critical region is split into 2.5% at each end. The result would have been significant if the P(sample mean > 103) < 0.025. The wording of the question is crucial.

Example: A machine produces bolts with a mean diameter (in cm) which is normally distributed such that $X \sim N(0.580, 0.015^2)$. After servicing, an operator suspects that the mean diameter has changed. A sample of 50 bolts has a mean diameter of 0.574 cm. Assuming the standard deviation is unchanged test, at the 1% level, whether this sample provides evidence that the mean diameter has changed.

Solution: We are not asked to test whether the diameter has decreased or increased, only whether it has changed – so we will carry out a 2-tailed test. Since the test is at the 1% (or 0.01) level, we therefore need to see if P($X < 0.574$) is less than 0.005.

H_0: $\mu = 0.580$

H_1: $\mu \neq 0.580$

Significance Level: 1%

2-tailed test

The GDC gives a p-value of 0.00468 < 0.01

Reject H_0. There is evidence to suggest that the mean diameter is no longer 0.580 cm.

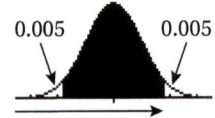

0.005   0.005

Finding the critical values: An alternative way of tackling the question above is to find the *critical* values; these are the boundary values where the "likely" region becomes the "unlikely" region. The unshaded areas under the normal curve illustrated represent the critical regions for a 2-tailed test with significance level 1%. The lower critical value can be found using the inverse norm function on your GDC (with area = 0.005, mean = 0.580 and SD = $\frac{0.015}{\sqrt{50}}$), and the upper critical value in the same way using an area of 0.995 (see arrow on the diagram). Check that these calculations give critical values of 0.57454 and 0.58546. The sample mean of 0.574 is therefore just inside the lower critical region, hence the significant result. This is also an illustration of using full accuracy.

I have used the standard error of the mean since in hypothesis testing we are always testing samples.

Unknown σ: On page 154 we saw how to calculate confidence intervals when the population standard deviation is unknown by using the t-distribution. So for a hypothesis test we use a t-test instead of a z-test.

Example: A data checker claims he can process database records at an average rate of 1 every 10 minutes. He is timed checking 5 records, and manages the following results (in minutes):

        9.88   10.18   10.23   10.39   10.25

Assuming the times are normally distributed, test at the 5% level whether his claim is likely.

Solution: H_0: $\mu = 10$

         H_1: $\mu \neq 10$ (Two tailed test because the bias could be either way)

         Sig. Lev. 5%

         Enter the data onto the GDC and the p-value is 0.0913. Since 0.0913 > 0.05 we accept H_0 and conclude that there is no evidence to suggest the mean is not 10 minutes.

Once again, we could use confidence intervals instead. Try entering the data from the previous question, then use the t-interval test as before, remembering to put in a confidence level of 0.95 (you would have put in 0.90 for a 1-tailed test. The confidence interval turns out to be [9.9526, 10.419]. The population mean, 10, is within these limits, and the result is therefore not significant.

Paired samples: In the previous section we were testing sample means against known population means to see if there was a significant difference. With a paired sample test, we are looking to see whether the mean for a particular sample has changed after an event. For example, suppose we test a sample of 10 people to see how many seconds they take to complete a task. After a week's training, we test them again to see if they are any quicker. The results could look like this:

	A	B	C	D	E	F	G	H	I	J
Before training	73	83	85	87	91	99	87	85	83	79
After training	73	79	81	86	87	91	84	83	84	76

The main point here is that we don't have two samples, just two sets of results for one sample; we are interested in the *differences* between the before and after results. So in effect we need to perform a t-test on the mean of the differences.

You could enter the results in two lists and then create a third list by subtraction. The test can then be carried out on the third list. Only worth it if there are a lot of values – or you're not very good at subtraction!

161

	A	B	C	D	E	F	G	H	I	J
Before training	73	83	85	87	91	99	87	85	83	79
After training	73	79	84	88	88	96	84	86	84	76
Differences	0	−4	−1	+1	−3	−3	−3	+1	+1	−3

The null hypothesis will be that the differences have a mean of 0.

H_0: $\mu_{\text{diff}} = 0$

H_1: $\mu_{\text{diff}} < 0$

SL: 5%

p-value = 0.0287 (GDC).

0.0287 < 0.05 so reject H_0. The training appears to have been successful.

But not at a significance level of 1%: 0.0287 > 0.01.

The *t*-test: The *t*-test is used to compare the means of two samples.

Case A: A strawberry grower believes that the plants in a field facing south have a greater yield that in another field which faces north. He weighs the strawberries on 20 plants in each field and calculates the mean weight in each case. A t-test can be used to test the hypothesis that the mean weight of strawberries in field 1 is greater than in field 2.

Case B: A group of boys and a group of girls each take a reaction time test. The researcher believes that the mean reaction time of the two groups will not be the same, but doesn't know which group might be quicker. A t-test can be used to test the hypothesis that the reaction time of the two groups is different.

 There are several options when carrying out a t-test on your calculator. In most cases you will choose:

- 2 sample test
- Data rather than stats
- Pooled (in every case)

You will need to enter:

- Data into the relevant lists
- The alternative hypothesis. In the two examples above, the case A alternative hypothesis is $\mu_1 > \mu_2$, and in case B it will be $\mu_1 \neq \mu_2$. A third option is $\mu_1 < \mu_2$.

When the results are shown, look for the p-value which can then be compared against the significance level – exactly the same as in the other types of testing.

One restriction: for the *t*-test to work, both samples must come from normally distributed populations. A question may ask you to test this; for example, by first carrying out a goodness of fit test, or examining the symmetry of a box plot.

Let's try the *t*-test at the 10% significance level on case B with the following data (in seconds):

Boys' reaction times: 0.34, 0.28, 0.36, 0.41, 0.21, 0.40

Girls' reaction times: 0.42, 0.44, 0.30, 0.29, 0.31, 0.38, 0.48

H_0: $\mu_1 = \mu_2$ H_1: $\mu_1 \neq \mu_2$

Note that in case A the null hypothesis would be $\mu_1 \leq \mu_2$.

On the face of it, with the boys' mean time of 0.33 s and the girls' mean time of 0.37 s, they do seem different. But you should find that the p-value is 0.353 which is way above 10%, so we accept H_0.

Two samples: Another type of test is where two samples are taken and we test whether or not they could come from normal distributions with the same mean (or non-normal if the sample is large enough). This test can also be one- or two-tailed depending on the wording of the question.

For example, suppose we have a survey from 2005 which shows that a random sample of 250 first time mothers had a mean age of 22.45 years and a standard deviation of 2.9 years. A similar survey was carried out in 2015 when 280 first-time mothers had a mean age of 22.96 years and a standard deviation of 2.8 years. Test at the 5% level whether or not these figures suggest the mean age of first-time mothers has changed.

H_0: $\mu_1 = \mu_2$ or "The mean age has not changed."

H_1: $\mu_1 \neq \mu_2$ or "The mean age has changed."

Enter the data into the GDC using a 2-sample t-test (because we don't know the population standard deviation). If your calculator shows "Pooled" choose "No."

The p-value is 0.0404 < 0.05 so we reject H_0. 10 years later it appears that the mean age of first-time mothers has changed.

Testing for the Mean: Practice Questions

1. A machine is designed to produce rope with a breaking strain of 1000 N. The standard deviation is known to be 21 N. A random sample of nine pieces of rope had a mean breaking strain of 1012 N.

    (a) Is there evidence at the 5% significance level that the breaking strain has changed?

    (b) Is there evidence at the 5% significance level that the breaking strain has increased?

2. A fruit buyer believes that the pineapples he buys from a particular grower are not as big as they used to be. He claims that the pineapples used to average 0.73 kg each. A random sample of 20 pineapples gives the following weights in kg:

    0.65   0.68   0.77   0.71   0.67   0.70   0.70   0.72   0.76   0.73

    0.75   0.69   0.72   0.73   0.69   0.71   0.70   0.69   0.75   0.78

    Carry out a test at the 5% significance level and state the conclusion, including any assumptions you have made.

3. A theme park manager counted the number of visitors to two separate attractions, "Manic Ride" and "Megasplash", over 8 days. The results are shown in the following table:

Day	A	B	C	D	E	F	G	H	I	J
Manic Ride	4122	1094	3082	1530	1854	2194	1396	3024	927	3763
Megasplash	3962	962	2824	1330	1552	2311	1260	2940	880	3821

Carry out a suitable test on the data to determine whether or not there is evidence that the Manic Ride is more popular.

Answers

1. (a) $p = 0.086$, accept H_0
   (b) $p = 0.043$, reject H_0

2. $p = 0.033$, reject H_0. The pineapples appear to be smaller than before. Assume weights form a normal distribution.

3. H_0: $\mu_{diff} = 0$
   H_1: $\mu_{diff} > 0$
   $p = 0.011$ so accept H_0
   There is no evidence that at the 5% level that Manic Ride is more popular.

4. Critical region is $\bar{X} > 13.14$ so accept H_0 since 13.0 < 13.14.

4. A sample is taken from a normal population having a $N(12.5, 1.5^2)$ distribution. Find the critical region for the test statistic $\bar{X}$ in the following test:

H_0: $\mu = 12.5$

H_1: $\mu > 12.5$

$n = 30$

Significance level: 1%

and hence determine whether a sample mean of 13.0 is significant.

4.22 Testing for Proportion using the Binomial Distribution

Suppose that the normal treatment for a particular medical condition has a probability 0.4 of success. A pharmaceutical company claims that a new drug increases the probability of success. In a clinical trial, a sample of 20 people with the condition were given the new drug, and 11 recovered from the condition. Let us test the hypothesis at the 5% level.

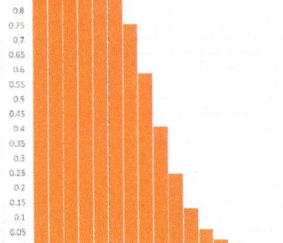

Using $X \sim B(20, 0.4)$, we find that $P(X \geq 11) = 1 - P(X \leq 10) = 0.1275$. Since $0.1275 > 0.05$ we conclude that this could easily have happened by chance, and there is no proof that the new treatment is any more effective than the old. Another way to tackle the problem is to find the critical region but, since we are dealing with a discrete distribution, it is unlikely that we will find a region with a probability of exactly 0.05.

This first chart shows the probabilities for $0 \leq X \leq 20$. We can see that the most likely outcome is 8, and at each end of the distribution the probabilities are too small to register.

The second chart shows the cumulative probabilities from the upper end. For example, $P(X \geq 9) \approx 0.4$. $P(X \geq 12)$ is just over 0.05, and $P(X \geq 13)$ is less than 0.05. (Check on your GDC: you should get 0.0565 and 0.0210). Thus the critical region at the 5% level is $X \geq 13$, and you would therefore need at least 13 patients out of 20 to recover to provide evidence that the new treatment has been successful – the critical value is 13.

Use trial and error to find the critical region at the 1% level. You should get $X \geq 14$.

4.23 Testing for the Mean of a Poisson distribution

The process for testing the mean of a Poisson distribution is very similar to testing the mean of a binomial distribution. Let's dive straight into a question to see it in action.

Road accidents occur on a sharp bend at a rate of 8 per month.

(a) Suggest a suitable probability distribution to model this situation.

The local council tries to reduce the accident rate by installing larger warning signs. In the month following the installation there were only 3 accidents. The council tell residents that the new signs have therefore been effective in reducing the accident rate.

(b) Test at the 5% level the council's claim.

The residents claim that the test should have been over a longer period, and suggest 3 months.

(c) Using a 5% significance level, find the critical region for the new test.

(a) $X \sim P_o(8)$

(b) $H_0: \lambda = 8$

   $H_1: \lambda < 8$

   Significance level = 5%

   $P(X \le 3) = 0.0423 < 0.05$

   $\therefore$ Reject H_0; there is evidence to support the council's claim

(c) $H_0: \lambda = 24$

   $H_1: \lambda < 24$

   Significance level = 5%

   $P(X \le 15) = 0.0344$, $P(X \le 16) = 0.0562$

   $\therefore$ Critical region is $X \le 15$

A one-tailed test is used because the council claims that the accident rate has been reduced.

You can speed up the trial and error calculations in part (c) by using a list instead of a single value of X. Another cunning way is to enter the cumulative Poisson distribution as a function (using the variable X instead of a value), and then using a table to see the results.

X	Y1
13	.01072
14	.01983
15	.0344
16	.05626
17	.08713
18	.12828
19	.18026

Y1 ■Poissoncdf(2...

Note that hypothesis testing on binomial and Poisson distributions will only be set as one-tailed, as will critical regions for those distributions.

4.24 Type I and Type II errors

We can never be certain of the conclusion when carrying out a significance test. It is possible that a test result leads us to reject H_0 when H_0 is in fact true. It is equivalent to a court passing a guilty sentence on an innocent defendant (who is presumed innocent unless proved otherwise). It is also possible that a test result leads us to accept H_0 when, in fact, H_1 is true - the guilty man being acquitted. The first of these errors is called *type I* and the second is called *type II*.

Type I errors are easy to deal with. If we choose a 5% significance level, this means that a significant result will happen *by chance alone* 5% of the time. Thus there is a 5% probability of a type I error occurring. For a continuous distribution (such as the normal) the probability of a type I error is the same as the significance level.

However, for discrete distributions, the probability of a type I error is the same as the probability of the critical region.

So why don't we make the significance level really small? Because the less chance there is of a type I error, the greater the chance of a type II error. Now, a type II error is rather

If we consider the example of testing a new treatment, then a type I error occurs when the number of people benefitting is so high that we conclude the treatment is working – when in fact it isn't. A type II error occurs when the number benefitting is no more than we would normally expect, but in fact the treatment is working.

Look back to page 164 and the example on testing for a proportion. The critical region was $X \ge 13$, which means that $P(\text{type I error}) = 0.021$.

more tricky. We want to know the probability that a result is ***not*** in the critical region if the population mean is actually different from what we think it is. But without knowing that mean, we can't do the calculation. And, if we knew the mean, there wouldn't be any point doing the significance test in the first place! The diagrams below may help to explain this contorted piece of reasoning.

Suppose we carry out a 5% significance test on the mean of a normal distribution $N(20, 2^2)$, and sample size = 10. Our test is that H_0: $\mu = 20$ and H_1: $\mu > 20$. The critical value (using an invNorm calculation) is 21.04 and hence the critical region is $\mu > 21.04$. So, for different ***actual*** population means, the shaded areas each show the probability that we will ***not*** get a significant result from our test – the Type II error. But this is very theoretical since we don't actually know the value of μ.

In practice, we would carry out more than one test to increase our chances of reaching the right conclusion.

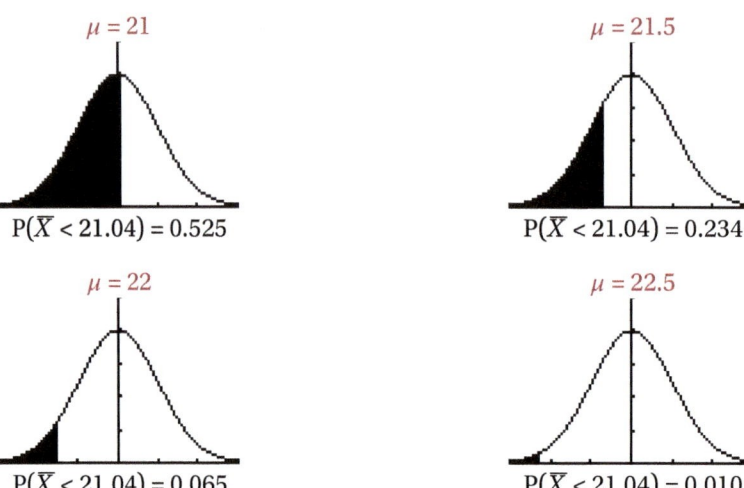

So let's see how this might appear in an exam question.

Boxes of breakfast cereal have contents whose masses should form a normal distribution $M \sim N(375, 152)$. It is thought that the machine which fills the boxes has a defect, and the mean is no longer 375 g. The contents of a random sample of 16 boxes are weighed and a significance test at the 5% level carried out.

(a) Write down null and alternative hypotheses.

This is a two-tailed test because we are testing whether the mean has changed.

H_0: $\mu = 375$

H_1: $\mu \neq 375$

(b) For what values of the sample mean is the alternative hypothesis accepted.

This is the same as finding the critical region. Because it is a two-tailed test, we use 2.5% at each end of the distribution.

The critical values are 367.65 and 382.35

H_1 will be accepted if $\overline{X} < 367.65$ or $\overline{X} > 382.35$.

(c) Given that the actual value of μ is 380 g, find the probability of making a type II error.

A type II error will be made if a result between the critical values is obtained. In other words, with a mean of 380, we must find the probability that $\overline{X}$ lies between the critical values.

With $N\left(380, \frac{15^2}{16}\right)$, $P(367.65 < \bar{X} < 382.35) = 0.734$

(d) How could the test be changed so as to reduce the probability of a type II error?

In fact, the probability is lowered to 0.621. Try it.

Increasing the significance level to 10% will increase the size of the critical region, and hence the probability of getting a sample mean within that region. A larger sample size would also reduce the probability of a Type II error.

Practice Exercise: Type I and Type II Errors

1. On a national basis, the number of people who pass their truck driving test first time is 40%. A driving school claims that their students do better than this because 25 of the last 50 passed first time. Use a significance test at the 5% level for a binomial distribution to test whether this claim is justified. Find the probability of a type II error if the proportion passing is in fact 45%.

2. The number of daily absences in a junior school can be modelled by a normal distribution with mean 1.94 and variance 0.357. A new system of incentives is devised to reduce absence and, in the first 40 days of the new system, it was found that there were 72 absences. Assuming the 40 days is a random sample:

    (a) Test at the 1% significance level whether the new system has reduced absences.

    (b) What is the probability of a Type I error in this test?

    (c) Calculate the probability of making a Type II error given that the mean number of absences is later found to be 1.88.

    (d) In the context of this questions, what is meant by Type I and Type II errors?

Answers

1. 9.8% > 5%, so reject H_0
   Critical value = 26
   $\mu = 0.45$,
   $P(X \le 25) = 0.803$

2. (a) .0066 < .01, reject H_0
   (b) 0.01
   (c) C.V. = 1.767,
   P(type II error) = 0.0217
   (d) Type I – conclude incentive system is working when it isn't; Type II – conclude incentive system isn't working when it is.

4.25 Goodness of fit tests

Uniform distribution: In the previous section the χ^2 test was used to test observed frequencies against expected frequencies in a contingency table. In a goodness of fit test, we are testing how well a single set of values fits a probability distribution. The simplest example is testing against a uniform distribution. For example, if a die is thrown 60 times, you would expect on average to get each of the values 1 to 6 ten times. You suspect that the die is biased so that 3 occurs more often than it should. Here are the observed results from 60 throws:

Number	1	2	3	4	5	6
Obs. freq.	9	10	18	8	7	8
Exp. freq.	10	10	10	10	10	10

Looks suspicious, doesn't it? Now enter the observed and expected frequencies into your GDC and apply the χ^2 goodness of fit test. You must enter the number of degrees of freedom which is always $n - 1$, where n is the number of possible outcomes. So in this case the number of degrees of freedom is 5. The p-value is 0.1456 > 0.05, so this distribution of results could have happened by chance.

Binomial and Poisson distributions: When testing a set of observed frequencies against either the Binomial or Poison distributions, the steps you take are the same as for the uniform distribution, but there are one or two differences in detail.

*   You may not know the value of p in the Binomial distribution. In such cases, you must estimate it from the given data. If you *do* have to estimate it, the number of degrees of

freedom will be **two** less than the number of frequencies; otherwise it will be one less. Exactly the same applies to the value of λ in the Poisson distribution.

- The Poisson distribution theoretically has an infinite number of values; in practice, lump together all those where the probability is pretty well 0, and find the combined probability by calculating (1 – all the others).

Note that for accurate results, each expected frequency must be greater than 5. If this isn't the case, you must combine columns.

Let's work through an example where we test a set of results to see if they are binomially distributed.

Example: A medical researcher wants to test whether a common condition has a genetic cause, and so runs in families, or whether it occurs independently with a fixed probability 0.4. She selects 200 sets of 6 closely related people and determines how many in each set have the condition. The results are shown in the following table:

Number	0	1	2	3	4	5	6
Obs. freq.	17	40	70	45	24	4	1

(a) If the expected results follow a binomial distribution with probability 0.4, work out the probability for the number of people in each set with the condition, and hence the expected frequencies for the 200 sets.

$X \sim B(6, 0.4)$. The probabilities are calculated using the Stats functions on the GDC and multiplied by 200 to give expected frequencies.

Number	0	1	2	3	4	5	6
Probability	0.0467	0.187	0.311	0.276	0.138	0.0369	0.0041
Exp. freq.	9.3	37.3	62.2	55.3	27.6	7.4	0.8

(b) Set up a hypothesis test and use the χ^2 goodness of fit test at the 5% level.

H_0: The underlying distribution is binomial with $p = 0.4$

H_1: The underlying distribution is not binomial

Number	0	1	2	3	4	5 & 6
Obs. freq.	17	40	70	45	24	5
Exp. freq.	9.3	37.3	62.2	55.3	27.6	8.2

The expected frequencies of the last two columns have been combined to ensure none are less than 5.

Degrees of freedom = 6 – 1 = 5

χ^2 statistic = 11.19

p-value = 0.0478

(c) What conclusion can be drawn?

Since 0.0478 < 0.05 reject H_0. The test supports the theory that the condition has a genetic cause.

Example: A commuter carries out a survey over 80 weeks to see how often his train was late over 5 working days. He wants to test the results to see if they follow a binomial distribution. Why might this assumption not be true? The results were:

Number of days	0	1	2	3	4	5
Frequency (of late trains)	10	14	31	16	8	1

Carry out a test at the 10% level to see if these results follow a Binomial distribution.

Solution: It may be that trains being late one day affect their punctuality the next day, so the probability of late trains each day may not be independent.

H_0: The Binomial distribution is a suitable model for the number of boys in these families.

H_1: The Binomial distribution is not a suitable model for the number of boys in these families.

There are 400 days in the survey, and on 161 days the train is late. So P(late) = 0.4025. To calculate the expected frequencies, we first work out the values of the binomial distribution B(5, 0.4025), then multiply them by 80.

> If you use a list {0,1,2,3,4,5} instead of the x value, your GDC will produce all the expected frequencies in one go, which you can multiply by 80. Or use a table of values

Number of boys	0	1	2	3	4	5
Observed frequency	10	14	31	16	8	1
Probability	0.0762	0.2565	0.3456	0.2328	0.0784	0.0106
Expected frequency	6.1	20.5	27.6	18.6	6.3	0.8

The last column has an expected frequency less than 5, so we combine it with the previous column.

Number of boys	0	1	2	3	4 - 5
Observed frequency	10	14	31	16	9
Expected frequency	6.1	20.5	27.6	18.6	7.1

$n = 5 - 2 = 3$ we subtract 2 because we have estimated P(late)

p-value = 0.0951

0.1194 > 0.1 so we accept H_0. The frequency of late trains does fit a binomial distribution.

Normal distribution: Testing a set of observed frequencies to see if they fit a Normal distribution requires the following points to be taken into account:

- The parameters μ and σ may be known. If not, they must be estimated from the data, remembering to use the formula shown on page 152 to calculate the unbiased estimate of population standard deviation.

- The Normal distribution is continuous, so your data must be collected into a frequency table. When finding the expected frequencies, the first and last classes in the table will be of the form $X < a$ and $X > b$ to mirror the fact that the Normal distribution ranges from $-\infty$ to $+\infty$ (see example below).

- The number of degrees of freedom is:
  - (Number of classes – 1) if μ and σ are known.
  - (Number of classes – 2) if you have to estimate *one* of the parameters.
  - (Number of classes – 3) if you have to estimate *both* parameters.

- Estimated frequencies are obtained from your GDC, and then multiplied by n, the number of values. They are not rounded to whole numbers.

- The usual constraint that classes with expected frequencies less than 5 must be combined with their neighbours.

Example: An ice-cream seller noted down how much ice-cream he sold during a 100 day period, and collected the information into a frequency table.

Ice cream sold per day (kg)	0–	10–	20–	30–	40–50
Number of days	10	26	43	14	7

It is thought that these values can be modelled by a normal distribution. Carry out a goodness of fit test using a 5% significance level.

We must first estimate the values of μ and σ, and this can best be done by entering the data onto a GDC: remember to use the mid-values of each class. $\mu = 23.2$, $\sigma_{n-1} = 10.29$. We can now find the expected frequencies for each interval by calculating the Normal probabilities and multiplying by 100.

Ice cream sold per day (kg)	<10	10–20	20–30	30–40	>40
Number of days (O)	10	26	43	14	7
Normal probability	0.0998	0.2781	0.3677	0.2031	0.0513
Expected number of days (E)	10.0	27.8	36.8	20.3	5.1

H_0: $N(23.2, 10.29^2)$ forms a suitable model for the data.

H_1: $N(23.2, 10.29^2)$ does not form a suitable model for the data.

$\chi^2 = 3.82$

$n = 5 - 3 = 2$

p-value $= 0.1478 > 0.05$

Accept H_0; the data could come from an $N(23.2, 10.29^2)$ distribution.

Practice Exercise: Goodness of Fit Tests

Answers

1. $p = 0.0323$, accept H_0

2. $p = 0.012$, reject H_0

3. (a) 0.4

   (b) $p = 0.015$, accept H_0

1. The diameters of a random sample of 30 machine-produced components were measured by passing them through a gauge. 21 passed through a 5 mm gauge but of these 13 failed to pass through a 3.5 mm gauge. Test at the 1% level the hypothesis that the components are manufactured with a mean diameter of 4.2 mm with a standard deviation of 0.8 mm.

2. A sample of 150 eggs is taken from the daily production of a poultry farmer and graded for size with the following results:

Weight (g)	45–	50–	55–	60–	65–
Frequency	12	29	67	32	10

Test whether this distribution fits a normal distribution with mean 58 g. Use a 5% significance level.

3. At a fairground stall, 6 rubber rings are thrown with the aim of landing them on a hook. The results for 100 people are shown in the table:

Successes	0	1	2	3	4	5	6
Frequency	0	26	36	20	10	6	2

(a) Estimate the probability of throwing a ring onto the hook.

(b) Test at the 1% level the hypothesis that the results are binomially distributed

Hint: Don't forget that observed frequencies cannot be less than 5.

4.26 Markov Chains

On page 100 we looked at an example of how a dynamic system transitioned to a steady state using the multiplication of a transition matrix. The same principle can be applied to probability.

Consider a student who travels to school by bus. If he misses the bus one day, there is a 0.6 probability he will miss is the next. But if he catches it, the probability he misses it the next day is 0.8 (obviously he became overconfident)! This information can be bundled up into a matrix P.

$$
\begin{array}{c}
\text{Second day} \\
\\
\end{array}
\begin{array}{cc}
& \text{First day} \\
& \begin{array}{cc} \text{Catches} & \text{Misses} \\ \text{bus} & \text{bus} \end{array} \\
\begin{array}{c} \text{Catches bus} \\ \text{Misses bus} \end{array} & \begin{pmatrix} 0.2 & 0.4 \\ 0.8 & 0.6 \end{pmatrix}
\end{array}
$$

This is a transition matrix, so if we find P^2 this will tell us the probabilities of him catching or missing the bus on the third day.

$$
\begin{array}{c}
\text{Third day} \\
\\
\end{array}
\begin{array}{cc}
& \text{First day} \\
& \begin{array}{cc} \text{Catches} & \text{Misses} \\ \text{bus} & \text{bus} \end{array} \\
\begin{array}{c} \text{Catches bus} \\ \text{Misses bus} \end{array} & \begin{pmatrix} 0.36 & 0.32 \\ 0.64 & 0.68 \end{pmatrix}
\end{array}
$$

Using the eigenvalue method on page 26 we find that, for large values of n, $P_n = \begin{pmatrix} \frac{1}{3} & \frac{1}{3} \\ \frac{2}{3} & \frac{2}{3} \end{pmatrix}$.

In other words, after many days it makes no difference what happened on the first day – he has a $\frac{1}{3}$ chance of catching the bus and a $\frac{2}{3}$ chance of missing the bus.

In a game, a line of people must whisper "Buy" or "Sell" to the next person. Each person either changes the message with fixed probability p, or repeats it. The initial matrix M showing the situation for the first two people is:

$$\begin{array}{cc} & \begin{array}{cc} \text{1st person} \\ \text{"Buy"} \quad \text{"Sell"} \end{array} \\ \text{2nd person} \begin{array}{c} \text{"Buy"} \\ \text{"Sell"} \end{array} & \begin{pmatrix} 1-p & p \\ p & 1-p \end{pmatrix} \end{array}$$

Given that $p = 0.1$,

    (a)   Calculate M^3, and explain what the first element of the matrix represents.

    (b)   Use the multiplication of three matrices to find M^n, and hence find the probability that the message has changed after passing down a long line.

(a) $\begin{pmatrix} 0.756 & 0.244 \\ 0.244 & 0.756 \end{pmatrix}$

There is a probability of 0.756 that if the first person said "buy", then the 4th person also said "buy".

(b) Eigenvalues are 0.8 and 1 with corresponding eigenvectors $\begin{pmatrix} -1 \\ 1 \end{pmatrix}$ and $\begin{pmatrix} 1 \\ 1 \end{pmatrix}$.

So $M^n = -\frac{1}{2}\begin{pmatrix} -1 & 1 \\ 1 & 1 \end{pmatrix}\begin{pmatrix} 0.8^n & 0 \\ 0 & 1 \end{pmatrix}\begin{pmatrix} 1 & -1 \\ -1 & -1 \end{pmatrix}$

$= -\frac{1}{2}\begin{pmatrix} -0.8^n - 1 & 0.8^n - 1 \\ 0.8^n - 1 & -0.8^n - 1 \end{pmatrix}$

As $n \to \infty$, $M^n \to \begin{pmatrix} 0.5 & 0.5 \\ 0.5 & 0.5 \end{pmatrix}$

P(message changing) = 0.5

As a matter of interest, it doesn't matter what value p has, in this situation the steady state matrix will always be the same.

Statistics and Probability: Long Answer Questions

Starting on the next page is a selection of Paper 2 style exam questions related to Statistics and Probability. The answers are given here, but full workings may be found on the Peak Study Resources website.

1. A group of 100 students in a school are asked about whether they study History or Biology. 10 study neither, 60 study Biology, 72 study History.

   (a) *n* students take both subjects.

      (i) Show that *n* = 42.

      (ii) Write down the number of students who only study Biology.

   (b) One student is selected at random.

      (i) Find the probability that the student studies only one of the two subjects.

      (ii) Given that the student only studies one subject, find the probability that he studies History.

   (c) Let *A* be the event that a student studies History, and *B* be the event that a student studies Biology.

      (i) Explain why *A* and *B* are not mutually exclusive.

      (ii) Show that *A* and *B* are not independent.

   (d) There are 380 girls and 420 boys in the school. An opinion poll is to be carried out using a sample of 40 students.

      (i) Why would a quota sample not give reliable results?

      (ii) How many boys and how many girls would be surveyed if a stratified sample were to be set up?

   _____

Answers:

   (a) (ii) 18

   (b) (i) $\frac{48}{100}$ (ii) $\frac{30}{48}$

   (c) (i) $A \cap B \neq \varnothing$ or "Some students study both subjects"

      (ii) $P(A) \times P(B) = 0.6 \times 0.72 = 0.432$. $P(A \text{ and } B) = 0.42$. $0.432 \neq 0.42$

   (d) (i) Quota sampling does not produce a random sample.

      (ii) 19 girls and 21 boys.

2. A test has five questions. To pass the test, at least three of the questions must be answered correctly.

The probability that Sammy answers a question correctly is $\frac{1}{4}$. If X is the number of questions that Sammy answers correctly,

(a) (i) Find $E(X)$ and $Var(X)$.

(ii) Find the probability that Sammy passes the test.

Martha also takes the test. Let Y be the number of questions that Martha answers correctly. The following table is the probability distribution for Y.

y	0	1	2	3	4	5
$P(Y = y)$	0.34	0.49	$p - 2q$	$p - 2q$	$p - 3q$	0.01

(b) (i) Show that $3p - 7q = 0.16$

(ii) Given that $E(Y) = 1$ find p and q.

(c) Find who is more likely to pass the test.

_____

Answers:

(a) (i) 1.25, 0.9375

(ii) 0.104

(b) (ii) $p = 0.1, q = 0.02$

(c) Martha

3. The scores for IQ tests are normally distributed with mean of 100 and standard deviation of 10. Paula gives an IQ test to 27 grade 12 IB students in her school. Her results are in the table below.

Score, x	Frequency
$x < 95$	1
$95 \leq x < 100$	4
$100 \leq x < 105$	6
$105 \leq x < 110$	7
$x \geq 110$	9

Paula wants to test if these results are taken from a population whose IQs follow the standard distribution of IQs as $N(100, 10^2)$, and performs a chi-squared goodness of fit test at the 1% significance level.

(a) Write down the null and alternative hypotheses.

(b) Complete the following table of expected values

Score, x	Frequency
$x < 95$	
$95 \leq x < 100$	5.17
$100 \leq x < 105$	5.17
$105 \leq x < 110$	
$x \geq 110$	4.28

(c) Combine rows to ensure all expected values are greater than 5 and give the new expected and observed tables.

(d) Perform the chi-squared test at a 1% significance level and state the conclusion for the test.

(e) Give two reasons why the test might indicate that the IQs of the students in the school do not follow a $N(100, 10^2)$ distribution.

Paula wants to use a t-test to see if there is evidence that the students in grade 12 have a higher average IQ than the grade 11 students.

(f) Explain why the result of her previous test might indicate that the t-test is not appropriate.

Answers:

(a) H_0: The distribution is $N(100, 10^2)$
H_1: The distribution is not $N(100, 10^2)$

(b)

Score, x	Frequency
$x < 95$	8.33
$95 \leq x < 100$	5.17
$100 \leq x < 105$	5.17
$x \geq 105$	8.33

(c)

Score, x	Observed	Expected
$x < 95$	1	8.33
$95 \leq x < 100$	4	5.17
$100 \leq x < 105$	6	5.17
$x \geq 105$	16	8.33

(d) p-value $= 0.0030 < 0.01$ so the result is significant at the 1% level and so the null hypothesis is rejected.

(e) For example:

– The school may be selective

– The sample might not be representative of grade 12 students.

(f) The t-test should only be used when the background population is normally distributed (or the sample is large).

4. The number of goals X scored in all football matches in local league one season was found to follow Poisson distribution such that $X \sim \mathrm{P_o}(2.4)$.

    Assuming the same distribution can be used for the following season, find

    (a) (i) The probability that a randomly chosen match ends with no goals scored

        (ii) The probability that 2 or more goals will be scored in a match

    (b) State a further assumption must be made in answering part (a)

    On one day, 12 league matches are played.

    (c) (i) State the distribution for the **total** number of goals scored in the 12 matches.

        (ii) Find the probability that at least 24 goals are scored in the 12 matches.

        (iii) Find the probability that at least 2 goals are scored in **each** of the matches.

        (iv) Explain why the answers to (ii) and (iii) are not the same.

    _____

    Answers:

    (a) (i) 0.0907

        (ii) 0.692

    (b) The number of goals scored in each match is independent of other matches.

    (c) (i) $\mathrm{P_o}(28.8)$

        (ii) 0.838

        (iii) Use $Y \sim \mathrm{B}(12, 0.692)$

            $P(Y = 12) = 0.0121$

        (iv) In (ii) the 24 or more goals can be scored in any of the matches, giving more possibilities than in (iii).

5. Over a period of time I record how long I have to wait for my bus every morning, and conclude that the waiting time, X, follows a distribution where $X \sim N(4.1, 1.7^2)$. When a new operator takes over the service, my waiting time over 10 random days is 3.2 minutes.

(a) Test at the 5% significance level whether the mean waiting time has decreased, stating the null and alternative hypotheses.

The new operator tries to improve the service further by increasing the frequency of the buses, thus reducing the queues at the bus stops. The time the bus has to wait at the 5 bus stops on my route before and after the change is shown in the following table:

Bus stop	A	B	C	D	E
Wait time before change (mins)	1.2	0.9	1.5	1.4	1.2
Wait time after change (mins)	1.0	0.8	1.4	1.0	0.9

(b) Determine if the mean wait time has reduced by carrying out a paired sample test at the 5% level.

My friend Alan often gets on the bus at bus stop C. If he does get on one day, then there is a probability 0.5 that he will also get on the next day. However, if he doesn't get on one day, then there is a probability of 0.8 that he won't get on the next day.

(c) (i) Write down a transition matrix to model this situation, labelling rows and columns appropriately.

(ii) If Alan gets on the bus on day 1, find the probability that he also gets on the bus on day 5.

(iii) Find the long-term probability that Alan gets on the bus on any given day, giving your answer as a fraction.

___

Answers:

(a) H_0: $\mu = 4.1$

H_1: $\mu < 4.1$

p-value $= 0.047 < 0.05$

Accept H_1. Mean waiting time has decreased.

(b) H_0: $\mu_{diff} = 0$

H_1: $\mu_{diff} < 0$

p-value $= 0.0837$

Accept H_0. The mean wait time has not reduced.

(c) (i)

$$
\begin{array}{cc}
 & \begin{array}{cc} \textit{Peter} & \textit{Peter not} \\ \textit{on bus} & \textit{on bus} \end{array} \\
\textit{Second day} \begin{array}{c} \textit{Peter on bus} \\ \textit{Peter not on bus} \end{array} & \left(\begin{array}{cc} 0.5 & 0.2 \\ 0.5 & 0.8 \end{array} \right)
\end{array}
$$

First day

(ii) 0.292

(iii) $\frac{2}{7}$

6. A school has been receiving complaints about its canteen so it decides to send out a questionnaire to a sample of their pupils.

The school has 58 boys and 52 girls in grade 10 and 1100 students in total. A sample of size 50 is to be selected and the sample will be stratified by grade and then by male/female within each grade.

(a) (i) Find the number of students from grade 10 who should be in the sample.

(ii) Write down how many of these students should be boys.

The questionnaire asks the pupils to give a score out of four to each of 20 categories with the higher mark indicating a greater level of satisfaction. Four weeks after the first questionnaire a second is given out to see if there has been an improvement.

To test for the reliability of the questionnaires eight students are selected and the average score given on the first questionnaire is compared with the score given on the second. The results are:

Student	A	B	C	D	E	F	G	H
First	3.5	2.4	2.7	3.2	1.9	2.9	3.6	2.1
Second	3.2	2.6	2.8	3.3	2.2	2.8	3.6	2.2

(b) (i) Find the correlation coefficient between the two sets of responses and comment on the reliability of the test.

(ii) State the name given to this type of reliability test.

In order to assess whether or not the satisfaction in each of the categories has improved the results from the whole sample of 50 students are looked at.

(c) State which test should be used to see if there is evidence of improvement in the level of satisfaction

Having looked at the results of the questionnaires the catering company announces that there has been an average improvement in 12 of the 20 categories.

(d) Assuming there has been no overall improvement in the level of satisfaction in the school, find the probability of at least 12 of the categories showing an improvement and comment on your result.

The catering company also states that in 2 of the 20 categories the improvement was such that there was significant evidence at the 5% level that an improvement had taken place.

(e) Assuming there has been no overall improvement in the level of satisfaction in the school find the probability of at least 2 out of 20 categories producing significant results at the 5% level by chance.

Comment on your results.

_____

Answers:

(a) (i) 5   (ii) 3

(b) (i) 0.970 = Strong evidence of reliability   (ii)   Test-retest

(c) Paired sample t-test

(d) 0.252

(e) 0.264. Neither result gives strong evidence of improvement

Chapter 5: CALCULUS

5.1 Differentiation – The Basics

Suppose we know that the rate of inflation is 3%. This fact is useful, but would be more useful if we knew how it was changing. If its rate of change is down 0.1%/month, we can make a guess at the rate of inflation in 6 months' time. Similarly, it is useful to know we are 100 km from our destination, even more useful if we know our rate of change of distance (ie speed) is 60 kmh⁻¹. The process of finding a "rate of change function" for a given function is called differentiation. You need to know the rules for differentiating different types of function, and for differentiating composite functions.

> The *gradient* of a graph at a point represents the rate of change of the function – so differentiation gives us the gradient of a graph at any point.

Notation: When you differentiate a function, the new function (the gradient function) is called the *derived* function (or *derivative*). If the original function is $f(x)$, the derived function is written as $f'(x)$. Alternatively, if the function is written in the form $y = f(x)$, the derived function is denoted by $\frac{dy}{dx}$.

> Don't confuse:
> $f'(x)$ Differentiated function
> $f^{-1}(x)$ Inverse function

Differentiating different types of function: You need to be able to differentiate various types of function (see table). If any functions are added or subtracted they can be differentiated independently. That is, $f(x) \pm g(x)$ differentiated is $f'(x) \pm g'(x)$.

This will not work for multiplication or division (eg to differentiate $(x + 1)(x - 2)$ you must first multiply out the brackets).

If a function is multiplied or divided by a *constant*, however, the constant just sits there: eg $2x^3$ differentiated is $2 \times 3x^2 = 6x^2$.

Also remember that functions of the form kx differentiate to give k, and that constants (which have a zero rate of change) differentiate to give 0.

$f(x)$	$f'(x)$
x^n	nx^{n-1}
$\sin x$	$\cos x$
$\cos x$	$-\sin x$
$\tan x$	$\sec^2 x$
e^x	e^x
$\ln(x)$	$\frac{1}{x}$

$x^2 - 3x$	$2x - 3$
$x^3 - 4$	$3x^2$
$2x(x - 1)$	$4x - 2$

Differentiating x^n: x^n differentiates to give nx^{n-1} for all $n \in \mathbb{R}$. This allows us to differentiate reciprocal and root functions. First, remember to write these functions as powers and with x in the numerator.

Examples are:

$f(x)$	$f(x)$ rewritten	$f'(x)$	$f'(x)$ simplified
$\sqrt{x}$	$x^{\frac{1}{2}}$	$\frac{1}{2}x^{-\frac{1}{2}}$	$\dfrac{1}{2\sqrt{x}}$
$\dfrac{4}{x^2}$	$4x^{-2}$	$-8x^{-3}$	$\dfrac{-8}{x^3}$
$x\sqrt{x}$	$x^{\frac{3}{2}}$	$\frac{3}{2}x^{\frac{1}{2}}$	$\dfrac{3\sqrt{x}}{2}$
$\dfrac{2}{\sqrt{x}}$	$2x^{-\frac{1}{2}}$	$-\frac{1}{2}\times 2x^{-\frac{3}{2}}$	$-\dfrac{1}{x^{\frac{3}{2}}}$

Differentiating $\sin x$, $\cos x$ and $\tan x$: x must be in radians for these differentiations to give correct results. eg: What is the gradient of the graph of $y = x + \sin x$ when $x = 1$? $\dfrac{dy}{dx} = 1 + \cos x$ so when $x = 1$, the gradient is $1 + \cos 1 = 1.54$. With the calculator set in degrees, you would get 1.9998.

Apart from adding or subtracting functions, we must use certain rules for differentiating functions when they are combined in different ways.

> Make sure you know how to use your GDC to find the gradient of a curve at a point.

5.2 The Chain Rule

The Chain Rule is used to differentiate composite functions. Consider the function $y = (4x + 3)^2$. If we write the "inner function" (ie $4x + 3$) as a single letter u, then the function becomes $y = u^2$. The chain rule shows us how the rates of change of **three** variables (as opposed to two) are connected:

$$\frac{dy}{dx} = \frac{dy}{du} \times \frac{du}{dx}$$

> I've written a fuller explanation of the Chain Rule on the website.

We can then use the chain rule like this:

$u = 4x + 3$ $\qquad\qquad \dfrac{du}{dx} = 4$

$y = u^2$ $\qquad\qquad\quad \dfrac{dy}{du} = 2u$

$$\frac{dy}{dx} = \frac{dy}{du} \times \frac{du}{dx} = 2u \times 4 = 8u = 8(4x + 3)$$

An alternative, informal, method is to proceed as follows:

- Take the "inner function" (in brackets) and differentiate it: 4
- Work out the "outer function" differentiated: $(\ldots)^2 \rightarrow \quad 2(\ldots)$
- Multiply the two together: $8(\ldots)$
- Fill in the brackets: $8(4x + 3)$

Here are more examples using the informal method:

$f(x) = \cos(2x - 4)$

> Inner function is $2x - 4$
>
> Differentiate inner $\rightarrow 2$
>
> Differentiate $\cos(...) \rightarrow -\sin(...)$
>
> Multiply $-2\sin(...)$

Result: $f'(x) = -2\sin(2x - 4)$

$f(x) = \ln(1 + x^2)$

> Inner function is $1 + x^2$
>
> Differentiate inner $\rightarrow 2x$
>
> Differentiate $\ln(...) = \dfrac{1}{(...)}$
>
> Multiply $2x \times \dfrac{1}{(...)}$

Result: $f'(x) = \dfrac{2x}{(1 + x^2)}$

$f(x) = \sqrt{1 - 5x} = (1 - 5x)^{\frac{1}{2}}$

> Inner function is $1 - 5x$
>
> Differentiate inner $\rightarrow -5$
>
> Differentiate $(...)^{\frac{1}{2}} \rightarrow \frac{1}{2}(...)^{-\frac{1}{2}}$
>
> Multiply $-\frac{5}{2}(...)^{-\frac{1}{2}}$

Result: $f'(x) = -\frac{5}{2}(1 - 5x)^{-\frac{1}{2}}$

$f(x) = e^{x^3} = e^{(x^3)}$

> Inner function is x^3
>
> Differentiate inner $\rightarrow 3x^2$
>
> Differentiate $e^{(...)} \rightarrow e^{(...)}$
>
> Multiply $3x^2 \times e^{(...)}$

Result: $f'(x) = 3x^2(e^{x^3})$

Chain Rule: Practice Exercise

1. Differentiate the following functions:

    (a) $\sin 2x$

    (b) $\ln(2x + 3)$

    (c) $\left(1 + \dfrac{1}{x}\right)^3$

    (d) $3\cos^2 x$

2. Given that $f(x) = e^{-x} + x$, find $f'(2)$ and the value of x for which $f'(x) = 0$.

3. Given that $f(x) = \dfrac{1}{x^2 - a}$ and $f'(2) = -1$, find the possible values of a.

Answers

1. (a) $2\cos 2x$

   (b) $\dfrac{2}{2x + 3}$

   (c) $\dfrac{-3}{x^2}\left(1 + \dfrac{1}{x}\right)^2$

   (d) $-6\sin x \cos x$

2. $-e^{-2} + 1$, $x = 0$

3. $a = 2$ or 6

Let $f(x) = \cos x$ and $g(x) = 2x^2$. Find expressions for $(g \circ f)(x)$ and $(f \circ g)'(x)$.

(a) $(g \circ f)(x) = g(\cos x) = 2(\cos x)^2 = 2\cos^2 x$

(b) $(f \circ g)(x) = f(2x^2) = \cos(2x^2)$

$u = 2x^2 \qquad \dfrac{du}{dx} = 4x$

$y = \cos u \qquad \dfrac{dy}{du} = -\sin u$

$\dfrac{dy}{dx} = \dfrac{dy}{du} \times \dfrac{du}{dx} = -4x \sin u = -4x \sin(2x^2)$

Calculus questions often involve maths from any of the other areas of the syllabus – in this case we have to find a composite function and then differentiate it.

We can identify it as a chain rule differentiation because one function is contained within a second function.

5.3 Product and Quotient Rules

When you have to differentiate two functions multiplied together you must use the *product rule*; and when two functions are divided, you must use the *quotient rule*. If the two functions are $u(x)$ and $v(x)$ – normally shortened to u and v – then the rules are:

- Product Rule: $$\frac{d}{dx}(uv) = u\frac{dv}{dx} + v\frac{du}{dx}$$

- Quotient Rule: $$\frac{d}{dx}\left(\frac{u}{v}\right) = \frac{v\frac{du}{dx} - u\frac{dv}{dx}}{v^2}$$

It may be helpful to think of the rules more informally as:

> **Product Rule:**
>
> (1st fn × 2nd fn differentiated) + (2nd fn × 1st fn differentiated)

Another quick way to remember them:

Product Rule is $uv' + vu'$

Quotient Rule is $\frac{vu' - uv'}{v^2}$

> **Quotient Rule:**
>
> $$\frac{\text{(bottom} \times \text{top differentiated)} - \text{(top} \times \text{bottom differentiated)}}{\text{(bottom line squared)}}$$

Note the plus sign in the product rule and the minus sign in the quotient rule. Also remember that, because of the minus sign, the order is important in the quotient rule.

When you are asked to do these more complicated differentiations, you can either write down every step in the formulae (safe but time-consuming) or you can do some of it in your head (faster, but you can go wrong). Here is an example of full working:

Example: Differentiate $y = x^2 \sin x$

Solution: $u = x^2 \quad \dfrac{du}{dx} = 2x$

$v = \sin x \quad \dfrac{dv}{dx} = \cos x$

$\dfrac{dy}{dx} = u\dfrac{dv}{dx} + v\dfrac{du}{dx} = x^2 \cos x + 2x \sin x$

It is possible that either (or both, if you are unlucky) of u and v are composite functions, in which case you will have to use the chain rule as well.

Example: Differentiate $f(x) = x^2 \ln(3x + 1)$

Solution: $u = x^2 \qquad\qquad \dfrac{du}{dx} = 2x$

$v = \ln(3x + 1) \qquad \dfrac{dv}{dx} = \dfrac{3}{3x + 1}$

$\dfrac{dy}{dx} = u\dfrac{dv}{dx} + v\dfrac{du}{dx}$

$\qquad = x^2 \dfrac{3}{3x + 1} + \ln(3x + 1) \times 2x$

$\qquad = \dfrac{3x^2}{3x + 1} + 2x \ln(3x + 1)$

Once you have differentiated, don't forget that the end result is, as with all differentiation, however complicated, the rate of change of the original function, the gradient of the graph at any point.

There is sometimes confusion when deciding whether to use the chain rule or the product rule. If x appears twice in the function, it's probably a product rule; if not, it's either a simple differentiation or it's the chain rule.

Product and Quotient Rules: Practice Exercise

1. Differentiate, simplifying where possible:

   (a) $x^3 \ln x$

   (b) $x \sqrt{x+3}$

   (c) $\sin x \cos(2x)$

   (d) $3x^2 e^x$

2. Differentiate, simplifying where possible:

   (a) $\dfrac{x+1}{2x-1}$

   (b) $\dfrac{x^2}{e^{2x}}$

   (c) $\dfrac{\sin x}{\cos x}$

Answers

1. (a) $x^2(1 + 3\ln x)$

   (b) $\dfrac{x}{2\sqrt{x+3}} + \sqrt{x+3}$

   (c)
   $-2\sin x \sin 2x + \cos x \cos 2x$

   (d) $3xe^x(x+2)$

2. (a) $\dfrac{-3}{(2x-1)^2}$

   (b) $\dfrac{2x(1-x)}{e^{2x}}$

   (c) $\dfrac{1}{\cos^2 x}$

5.4 Second Derivative

Notation: When a function is differentiated a second time, use the notation $\dfrac{d^2y}{dx^2}$ or $f''(x)$.

Interpretation: The first derivative gives us the gradient function, so the second derivative gives us the "rate of change of gradient" function. If, for example, $f''(3) = 2$, this means that when $x = 3$, the gradient of the graph is increasing at a rate of 2 (for every increase in x of 1). It does not necessarily mean that the gradient itself is positive – only that it is increasing. This tells us about the shape of the curve. The diagram below shows what happens for various values of the first and second derivatives and covers every possible point on any curve.

	$\dfrac{dy}{dx} < 0$	$\dfrac{dy}{dx} = 0$	$\dfrac{dy}{dx} > 0$
$\dfrac{d^2y}{dx^2} < 0$		MAXIMUM	
$\dfrac{d^2y}{dx^2} = 0$		POINTS OF INFLEXION	
$\dfrac{d^2y}{dx^2} > 0$		MINIMUM	

Imagine the graph is a road, and you are driving from left to right.

Right hand bends represent a decreasing gradient, so the second derivative < 0. Left hand bends represent an increasing gradient, so the second derivative > 0.

Points of inflexion occur whenever the steering wheel is momentarily straight: second derivative = 0 but this *doesn't* have to be when the gradient of the graph is 0.

Note that the parts of curves in the top row are known as "concave down" and those in the bottom row as "concave up."

It is important to remember the following:

- For a point of inflexion to occur $f''(x) = 0$, but the gradient at a point of inflexion is not necessarily 0.
- A point where $f''(x) = 0$ is not necessarily a point of inflexion. For example, $y = x^4$ has a **minimum** when $f''(x) = 0$.
- The sign of the second derivative at a turning point identifies the nature of the point: a maximum if $f''(x) < 0$, a minimum if $f''(x) > 0$.

Stationary points: A *stationary point* is one where the gradient is zero. If it is a local maximum or a local minimum, it can also be called a *turning point*. It isn't enough just to be able to find turning points using your GDC, as the next example shows. You must be able to differentiate a function and solve $f'(x) = 0$. Furthermore, you need to have a good understanding of the relationship between the curves of $f(x)$, $f'(x)$ and $f''(x)$ as the following section on the graphical behaviour of functions illustrates.

Marketing tests carried out on a new product suggest that the "desirability factor" is strongly influenced by price, and the model $f(x) = xe^{-x}$ is proposed where x is the price in dollars, and f the desirability factor, which is 0 for "definitely wouldn't buy" and 1 for "definitely would buy."

(a)   Find the desirability factor when the product price is set at \$2.

(b)   Find:

   (i) $f'(x)$

   (ii) $f''(x)$

(c)   Use your answer to (b)(i) to determine the stationary point on the curve of f, and your answer to (b)(ii) to demonstrate that it is a maximum.

(d)   Interpret your answers to (c) in terms of the model.

(a) $x = 2, f(x) = 2e^{-2} = 0.271$

(b)   (i) $u = x$ $u' = 1$

          $v = e^{-x}$ $v' = -e^{x}$

         $f'(x) = uv' + vu'$

            $= -xe^{-x} + e^{-x}$

    (ii) $f''(x) = -(-xe^{-x} + e^{-x}) - e^{-x} = xe^{-x} - 2e^{-x}$

(b) We must use the product rule to differentiate.

To find the second derivative, note that the first term in $f'(x)$ is exactly the same as $f(x)$, with a minus sign. We've already differentiated that! Also note that we don't necessarily have to find the value of $f''(x)$ – just its sign.

(c) For a stationary point $f'(x) = 0$

$$-x e^{-x} + e^{-x} = 0$$

$$e^{-x}(1 - x) = 0$$

$$x = 1$$

When $x = 1$, $y = 0.368$
So the stationary point is $(1, 0.368)$

When $x = 1$, $f''(x) = -0.368 < 0$
So the stationary point is a maximum.

(d) The optimum price for the product is $1 since this gives the greatest desirability factor.

5.5 Graphical Behaviour of Functions

If we are shown the graph of a function we can tell where the values of $f(x), f'(x)$, and $f''(x)$ are increasing, decreasing or zero. Slightly harder, if we are given the graph of $f'(x)$ we can do the same thing. The key point is that we do not need to know what the actual function is.

For example, suppose this sketch is part of the graph of $f'(x)$:

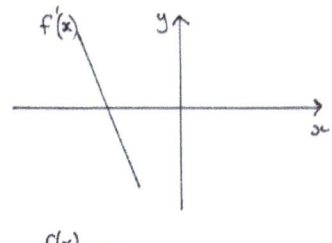

$f'(x)$ tells us the *values* of the gradient of $f(x)$. So when $f'(x) > 0$ (ie its graph is above the x-axis) the graph of the function has positive gradients – it is an *increasing* function. When $f'(x) = 0$ we've got a stationary point; and then the values of the gradient go negative, so we have a *decreasing* function.

What we don't know are the actual values of the function, only its behaviour. So the next sketch shows what the graph of the function is doing, but at this stage I can't put any axes in.

A couple of examples should cover the various techniques you will need to answer exam questions.

Example: The following diagram shows a part of the graph of $y = f(x)$.

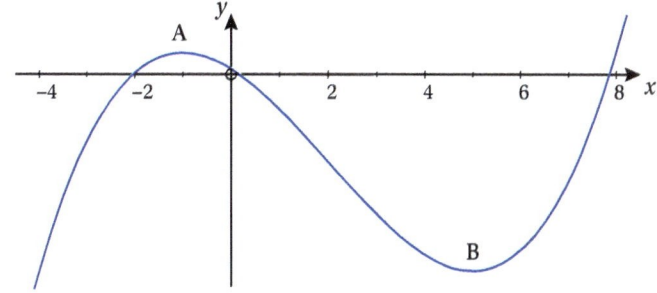

The graph has a local maximum at A, where $x = -1$, and a local minimum at B, where $x = 5$.

(a) For what values of x is f a decreasing function?

(b)   Sketch the graph of $f'(x)$.

(c)   Write down the following in order from least to greatest:

$f(0), f'(5), f''(-1)$.

Solution: (a) $-1 < x < 5$. (It is ***not*** decreasing at the turning points themselves).

(b)   To draw the sketch, let's consider the values of $f'(x)$. Up to the point A, f is an increasing function, so the values of f' are positive – but getting less so. At A the gradient is 0, and then it is negative until B. Between A and B there is a point of inflexion; at this point the gradient reaches its largest negative value. After B the gradient is positive again, and increasing. This leads to the following sketch:

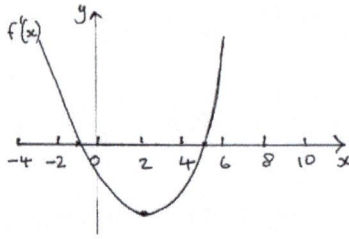

How do I know it's a curve? Because if it were two straight lines this would lead to a sharp point at $x = 2$, and this can't happen.

(c) $f(0)$ is positive, from original graph.

$f'(5) = 0$ because it's a stationary point

$f''(-1)$ is negative, because it's a maximum point

So, in increasing order, $f''(0), f'(5), f(0)$

Now have a look at this graph which shows the derived function, $f'(x)$, of a function which has a domain $-3 \le x \le 3$.

The graph of $f'(x)$ has minimum points when $x = a$ and when $x = c$, and a maximum point when $x = b$. It has zeroes when $x = a$ and $x = d$.

What can we deduce about the function?

- When is the function decreasing? This occurs whenever the gradient is negative, so when the graph of f' is below the axis. Answer: $0 < x < d$

- Where does the graph of $f(x)$ have a minimum? This will occur when the gradient is 0, so it could be at $x = a$, $x = 0$, or $x = d$. But a minimum has positive gradient to the left and negative gradient to the right; looking at the ***values*** of $f'(x)$ we can see that the only point which fits the criteria is where $x = 0$.

- Points of inflexion occur when $f''(x) = 0$, so at stationary points on $f'(x)$. At the same time, the gradient must have the same sign either side of the point of inflexion (ie both positive or both negative). Thus all of $x = a$, $x = b$, and $x = c$ are points of inflexion on the graph of $f(x)$.

I did this sketch in pen so it would reproduce properly. But I suggest you ***always*** use pencil for your diagrams and graphs. So much easier to make corrections.

5.6 Tangents and Normals

Tangents: Since differentiation gives us a way to calculate the gradient of a curve at any point, we can extend the calculation to work out the equation of the tangent to a curve at any point. On page 44 you will find the methods for finding the equation of a straight line given a point and the gradient, and this gives us all the tools we need for the next question.

The diagram shows a sketch of the graph of $y = x^2 - 4x + 3$.

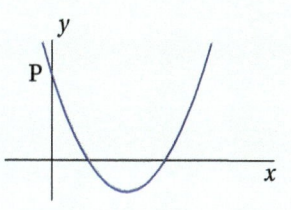

(a) Write down the co-ordinates of the point P where the graph crosses the y-axis.

(b) Show that the equation of the tangent to the point P is $y + 4x = 3$.

(a) When $x = 0$, $y = 0 - 0 + 3$.

So the coordinates of P are (0,3)

(b) $\dfrac{dy}{dx} = 2x - 4$

When $x = 0$, $\dfrac{dy}{dx} = -4$

$y - y_1 = m(x - x_1)$

$y - 3 = -4(x - 0)$

$y - 3 = -4x$

$y + 4x = 3$

In part (b) you could also have found the gradient at $x = 0$ using the GDC. However, if the question had first asked you to differentiate the function then the answer would have looked like this.

Normals: A tangent to a curve at a point has the same gradient as the curve at that point; a *normal* is perpendicular to the curve at that point. The procedure for finding the equation of the normal is exactly the same as for the tangent, except that we need to use the formula on page 44 for finding the gradient of one line which is perpendicular to another.

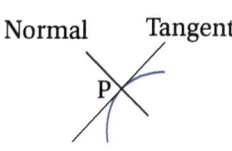

So, in the question above, the gradient of the normal at P is $\frac{1}{4}$; check that this gives the equation of the normal as $4y - x = 12$.

Let $y = 4 - \dfrac{a}{x}$, where $a \in \mathbb{R}$.

(a) Differentiate y with respect to x.

It is given that the normal to y at the point where $x = -2$ has a gradient of 1.

(b) Find the value of a.

(a) $\dfrac{dy}{dx} = \dfrac{a}{x^2}$

(b) $a = -4$

Using the GDC: Find out how to draw the tangent to a curve on your GDC; when you draw the tangent, you should find that the equation will be displayed as well: see the screenshot which displays the tangent to $y = x^2 - 3x - 1$ at the point (2, –3). Note that if you draw the normal to the curve at the same point, it will only look perpendicular to the curve if you are using the equal scales view.

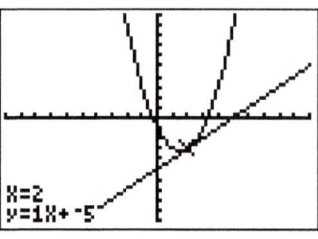

Tangents and Normals: Practice Exercise

Answers

1. (a) $y = 8x + 6$;
   $x + 8y + 17 = 0$
   (b) $y = x - 3$; $x + y = 1$
2. (2, 11)
3. (a) $a = 3$ (b) $y = 3x - 3$
   (c) $x + 3y = 11$ (d)

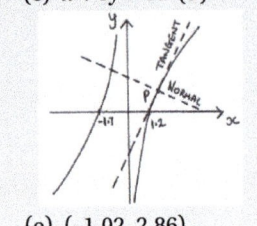

   (e) (–1.02, 2,86)

1. Find both the tangent and the normal to:

   (a) $y = 2x^3 - x^2 + 1$ at the point (–1, –2)

   (b) $y = -\dfrac{4}{x^2}$ at the point (2, –1)

2. T is the tangent to the curve $y = x^2 + 6x - 4$ at (1, 3) and N is the normal to the curve $y = x^2 - 6x + 18$ at (4, 10). Find the point of intersection of T and N.

3. For the function $f(x) = 2x - \dfrac{4}{x} + 1$

   (a) The point P = (2, a) lies on the graph of f. Find the value of a.

   (b) Find the equation of the tangent to the graph at P.

   (c) Find the equation of the normal to the graph at P.

   (d) Sketch the graph for values of x between –4 and 3. On the same graph, sketch the tangent and normal lines found in parts (b) and (c).

   (e) Where does the normal cut the curve again?

5.7 Optimisation problems

The point where a graph "turns round" can be very significant. For example, if the graph shows values of profit against selling price for a particular product, the maximum shows the selling price which leads to maximum profit.

We have seen that the turning points on a graph, known as either minimums or maximums, can be found by solving $f'(x) = 0$. Optimisation problems boil down to this:

- Find a function relating two real-life quantities
- Find the value of the first variable which leads to a maximum or minimum (the *optimum value*) of the second variable

The second part is just the same as finding a turning point on a graph – it's often the first part which causes head scratching because it involves setting up an equation.

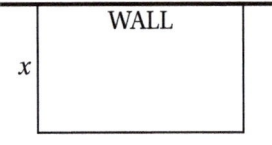

Let us suppose that a farmer has 80 m of fencing which he is going to use to enclose a rectangular sheep pen – he's going to use a stone wall as one side, and the fence to make up the other three sides of rectangle. What dimensions should the rectangle be so as to enclose the maximum area?

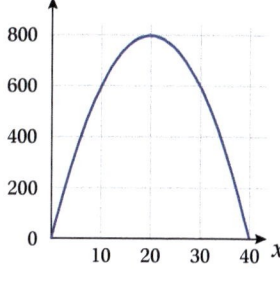

Let's call the width of the rectangle x. The opposite side will also be x leaving $80 - 2x$ for the length. So the area is $x(80 - 2x)$. Now it's the area we want to optimise, so:

$$A = x(80 - 2x)$$

$$= 80x - 2x^2$$

$$\frac{\mathrm{d}A}{\mathrm{d}x} = 80 - 4x$$

For a maximum, $80 - 4x = 0 \Rightarrow x = 20$.

So the dimensions of the rectangle are $20\,\text{m} \times 40\,\text{m}$, giving an area of $800\,\text{m}^2$. We don't need it, but I've also plotted the graph of $A = x(80 - 2x)$ to show that there is a maximum at $(20, 800)$.

Here's a harder one: what is the minimum distance from the point P(3, 1) to the curve $y = x^2$?

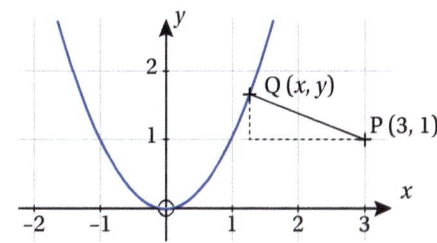

The diagram shows that we need the perpendicular distance from P to Q, but we don't know the coordinates of Q. But, using Pythagoras' Theorem, we can use the right angled triangle to get the following expression for the distance PQ: $\sqrt{(x-3)^2 + (y-1)^2}$.

We also know that, since Q is on the curve, $y = x^2$. We can therefore write the distance PQ just in terms of x: $\sqrt{(x-3)^2 + (x^2 - 1)^2}$. We don't know how to differentiate this to find a minimum, but we can enter it as a function on the GDC and find the minimum that way. This gives the minimum distance when $x = 1.289$. But be careful: the y value on the GDC is not the y-coordinate of Q, but the minimum distance, since we have entered the function which calculates the distance to the curve from P. And the minimum distance is therefore 1.834. If you need the y-coordinate of Q, just square x; Q is $(1.289, 1.663)$

Another method could be to find the normal to the curve which passes through P; then find where that normal intersects the curve, which will be Q; then find the distance PQ. Perhaps you might like to try it?

Example: A ship is to make a voyage of $200\,\text{km}$ travelling at a constant speed. When this speed is $v\,\text{km/h}$, the cost is $\$\left(v^2 + \frac{16\,000}{v}\right)$ per hour. Find the speed at which the ship should travel to minimise this cost. Find also the time taken for the voyage, and the cost.

Solution: Differentiating the expression for the cost we get $2v - \frac{16\,000}{v^2}$, and this must equal 0 for a minimum. This gives $v = 20\,\text{km/h}$. At $20\,\text{km/h}$ the voyage of $200\,\text{km}$ takes 10 hours. The cost is \$1200 per hour, giving a total cost of \$12\,000.

189

A rectangular sheet of cardboard with dimensions 24 cm by 15 cm has four squares cut off at each corner. The squares each have side x cm.

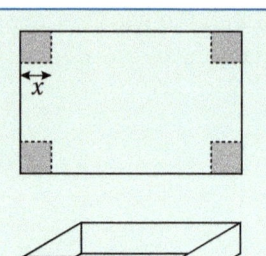

The remaining cardboard is then folded up to form a tray as shown in the second diagram.

(a) Write down, in terms of x, the length and width of the tray.

(b) Hence show that the volume of the tray, V, is given by $V = 4x^3 - 78x^2 + 360x$

(c) Find $\dfrac{dV}{dx}$

(d) Hence find the value of x which makes the volume a maximum, and calculate the maximum volume.

(a) Length $= 24 - 2x$

Width $= 15 - 2x$

(b) $V = x(24 - 2x)(15 - 2x)$

$= x(360 - 78x + 4x^2)$

$= 4x^3 - 78x^2 + 360x$

(c) $\dfrac{dV}{dx} = 12x^2 - 156x + 360$

(d) Maximum when $12x^2 - 156x + 360 = 0$

$x = 3$ or 10

10 cm isn't possible since the squares would be longer than the original sides.

So $x = 3$ cm and $V = 3 \times 18 \times 9 = 486$ cm^3

Part (b) is a "show that" question. Make sure you write down enough working to show you get to the solution.

Part (d) starts with "Hence". This means you must use the result of part (c) and not just draw a graph of V against x to find the maximum. And note also that this is another instance where only one of the solutions of the quadratic equation is valid in this particular context.

5.8 Related Rates of Change

If Anna runs twice as fast as Berta, and Berta runs three times faster than Carla, then Anna runs six times faster that Carla. That, in essence, is what related rates of change is about, and comes directly from the chain rule: $\dfrac{dy}{dx} = \dfrac{dy}{du} \times \dfrac{du}{dx}$. It comes in useful when we are dealing with rates of change which involve three variables. Here's a typical example:

Example: If the radius of a sphere is increasing at 3 cm s^{-1}, find the rate at which the volume is increasing when the radius is 4.5 cm.

Solution: First, identify the three variables – these are V, r, t. Next, state in terms of the variables what we know and what we want to know. We know that $\dfrac{dr}{dt} = 3$ and that we want to find $\dfrac{dV}{dt}$. Now write down the chain rule, starting with what you want to find: $\dfrac{dV}{dt} = \dfrac{dV}{dr} \times \dfrac{dr}{dt}$. We now need a connection between V and r,

Note that you can imagine that the dr cancels top and bottom, leaving you with $\dfrac{dV}{dt}$.

and this comes from the formula for the volume of a sphere $V = \frac{4}{3}\pi r^3$ which we can differentiate to get $\frac{dV}{dr} = 4\pi r^2$.

Finally, put all the information together in the chain rule: $\frac{dV}{dt} = 4\pi r^2 \times 3 = 12\pi r^2$. When $r = 4.5$, V is increasing at $763.4 \, \text{cm}^3\text{s}^{-1}$.

Sometimes a question may not be quite so straightforward, but the method should always be the same. For example, the diagram shows sand being poured to form a cone at a rate of $200 \, \text{cm}^3 \text{min}^{-1}$. Given that the height and radius are always equal, find the rate of increase of height in cm min^{-1} when the height of the cone is $10 \, \text{cm}$.

On the face of it we have 4 variables (V, h, r, t) but we know that $h = r$. We can therefore express the volume of the cone just in terms of h. Now, we want $\frac{dh}{dt}$ so we can write the chain rule as $\frac{dh}{dt} = \frac{dh}{dV} \times \frac{dV}{dt}$. Note that having written V in terms of h and differentiated, just take the reciprocal to find $\frac{dh}{dV}$. The answer is $0.637 \, \text{cm min}^{-1}$.

Full working is on the website

5.9 Kinematics (1)

Velocity and acceleration: Since velocity is rate of change of displacement, differentiating a displacement-time function will give velocity. Similarly, differentiating a velocity-time function will give acceleration (which is the rate at which velocity changes).

Let's consider the motion of a ball thrown straight up in the air and whose height h m at time t seconds is given by $h = 20t - 5t^2$, for $0 \leq t \leq 4$. We can find its velocity at any time t by differentiating: $v = \frac{dh}{dt} = 20 - 5t$. The following table shows its height and velocity at different times:

t (seconds)	0	1	2	3	4
h (m)	0	15	20	15	0
v (m/s)	20	10	0	−10	−20

What does this tell us? Looking at the height, it appears to reach a maximum of $20 \, \text{m}$ before falling down and hitting the ground at $4 \, \text{s}$. The initial velocity is $20 \, \text{m/s}$; at $2 \, \text{s}$ its velocity is $0 \, \text{m/s}$, confirming we are at maximum height, and then the velocity becomes negative, showing that it has reversed direction.

In the previous section we saw that to find a maximum or minimum we differentiate and set equal to 0. This is exactly what we have done here, the only difference being that the differentiated function actually has a physical meaning – velocity.

An interesting point is the difference between *displacement* and *distance*. At $t = 3 \, \text{s}$, the ball's displacement is $15 \, \text{m}$; this is the difference between its current position and its initial position. But the distance it has travelled is $20 + 5 = 25 \, \text{m}$. We shall look at this again on page 195 in the integration section.

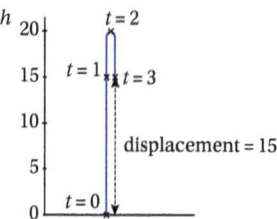

5.10 Indefinite Integrals

Integration is sometimes called "anti-differentiation": that is, it is the reverse operation to differentiation. However, the notation is very different, and you must understand two forms – the indefinite and the definite integral.

Notation: If we just consider functions of the form ax^n, then to reverse the differentiation process we must add 1 to n then divide by the new power. For example, $4x^2$ integrated is $\frac{4}{3}x^3$. The full notation for this is: $\int 4x^2\,dx = \frac{4}{3}x^3$. The $\int$ sign means "integrate", then comes the function you want to integrate, then dx (which shows you are "differentiating with respect to x"). However, the answer is not entirely correct. If you differentiate $\frac{4}{3}x^3$ you will certainly get $4x^2$, but this will also be true if you differentiate $\frac{4}{3}x^3 + 2$, $\frac{4}{3}x^3 - 1$, and so on. In other words, when we integrate, there could be a constant at the end. Since we don't know what it is, we add a 'c' which is called "the constant of integration." So we end up with $\int 4x^2\,dx = \frac{4}{3}x^3 + c$, and you must remember to add c to every indefinite integral – hence the word "indefinite."

Integrating x^n: Generally, $\int ax^n\,dx = \dfrac{ax^{n+1}}{n+1} + c$ and, as with differentiation, $n \in \mathbb{Q}$.

There is one exception, and that is when integrating $\frac{1}{x}$. Since this is x^{-1}, the rule above would give $\frac{x^0}{0}$ and this is undefined. But since differentiating $\ln x$ gives $\frac{1}{x}$, it follows that $\int \frac{1}{x}\,dx = \ln|x| + c$.

> The absolute value is required so that we can deal with negative values of x.

Integrating other functions:

$f(x)$	$\int f(x)\,dx$
$\sin x$	$-\cos x$
$\cos x$	$\sin x$
$\dfrac{1}{\cos^2 x}$	$\tan x$
e^x	e^x

Also, as with differentiation, it is true that

$$\int f(x) + g(x)\,dx = \int f(x)\,dx + \int g(x)\,dx \text{ and } \int kf(x)\,dx = k\int f(x)\,dx$$

Integrating functions of the form $f(ax + b)$: Consider what you get when you differentiate $f(x) = \sin(2x + 3)$. The "inner function" (see Chain Rule on page 180) is $2x + 3$, and this differentiates to give 2, so overall we get $f'(x) = 2\cos(2x + 3)$. Now let's reverse the process to find $\int \cos(2x + 3)\,dx$. Since $2\cos(2x + 3)$ must integrate to give $\sin(2x + 3)$, it follows that $\int \cos(2x + 3)\,dx = \frac{1}{2}\sin(2x + 3) + c$. This leads to the general result: $\int f(ax + b)\,dx = \frac{1}{a}F(ax + b) + c$, where $F(x) = \int f(x)\,dx$.

Here are some more examples:

$$\int e^{3x-1}\,dx = \tfrac{1}{3}e^{3x-1}$$

$$\int \frac{1}{2x-4}\,dx = \tfrac{1}{2}\ln(2x - 4) + c$$

$$\int (3 - 4x)^2\,dx = -\tfrac{1}{4} \times \tfrac{1}{3}(3 - 4x)^3 = -\tfrac{1}{12}(3 - 4x)^3 + c$$

If there is a constant multiplying the function, just leave it sitting around whilst you do the integration – it plays no part in the proceedings, but might help simplify the final result.

For example:

$$\int 4\sin(2x+1)\,\mathrm{d}x = 4\times(-\tfrac{1}{2})\cos(2x+1) = -2\cos(2x+1)+c$$

Reversing the chain rule: Integrating $f(ax+b)$ is a specific instance of integration by reverse chain rule. More generally, if you spot an integral which is made up of a composite function multiplied by the inner function differentiated, you can guess the solution and see what happens when you differentiate it. Putting it in words is pretty incomprehensible so perhaps a diagram will help:

$$\int 3x^2\cos(x^3)\,\mathrm{d}x$$

This function is this one differentiated,
so let's first try differentiating $\sin(x^3)$.

Using the chain rule, we find that $f(x) = \sin(x^3) \Rightarrow f'(x) = 3x^2\cos(x^3)$ which is exactly what we want. So $\int 3x^2\cos(x^3)\,\mathrm{d}x = \sin(x^3)+c$.

Sometimes you have to adjust by multiplying by a constant. For example, let's examine $\int 4x(x^2+3)^3\,\mathrm{d}x$. The overall function is a cubic, so will integrate to power 4. The inner function is x^2+3 and that differentiates to give $2x$; well, we've got $4x$ in front of the bracket, so a multiplier will sort that out. One way to tackle this is to try differentiating $(x^2+3)^4$, and this yields $8x(x^2+3)^3$. Comparing with the required integral, we conclude that $\int 4x(x^2+3)^3\,\mathrm{d}x = \tfrac{1}{2}(x^2+3)^4+c$.

> In all these cases it is also possible to use integration by substitution.
>
> You'll find the full working for this one on the website.

Here are some $f(ax+b)$ and reverse chain rule examples for you to try:

(a) $\int\frac{1}{4x-3}\,\mathrm{d}x$;   (b) $\int e^{2x}\,\mathrm{d}x$;   (c) $\int 4x\sin(x^2)\,\mathrm{d}x$;   (d) $\int 9x^2\sqrt{x^3-4}\,\mathrm{d}x$

(e) $\int\frac{6x^2}{x^3-2}\,\mathrm{d}x$ *(Try ln (x^3-2))*;   (f) $\int\frac{1}{\sqrt{2x-1}}\,\mathrm{d}x$;   (g) $\int\cos x(\sin^3 x)\,\mathrm{d}x$

Answers
(a) $\tfrac{1}{4}\ln(4x-3)+c$
(b) $\tfrac{1}{2}e^{2x}+c$
(c) $-2\cos(x^2)+c$
(d) $2(x^3-4)^{\frac{3}{2}}+c$
(e) $2\ln(x^3-2)+c$
(f) $\sqrt{2x-1}+c$
(g) $\tfrac{1}{4}\sin^4 x+c$

Solving gradient function equations: In some questions we are given the gradient (ie derived) function and asked to find the original function which gave rise to it. This means we must integrate the gradient function.

For example, if $f'(x) = 3x^2-x^3$, find $f(x)$.

So, $f(x) = \int 3x^2-x^3\,\mathrm{d}x = x^3-\tfrac{1}{4}x^4+c$, but we will need more information to find the value of c.

Suppose we know that $f(2)=6$ (in other words, when $x=2$, $y=6$). We then substitute this into the equation to get: $6 = 8-4+c$, and $c=2$. So the function we are looking for is $f(x) = x^3-\tfrac{1}{4}x^4+2$.

Example: It is known that for a function f, $f'(x) = ax-6$. Given that at the point on the graph of f where $x=1$, the y-coordinate is –1 and the gradient is –2, find the value of a and hence the function f.

Solution: The gradient at any point on the graph is $ax-6$. When $x=1$, the gradient is –2, so $a-6=-2 \Rightarrow a=4$.

So $f'(x) = 4x-6 \Rightarrow f(x) = 2x^2-6x+c$

When $x=1$, $f(x) = -1$

$$-1 = 2 - 6 + c \quad \Rightarrow \quad c = 3$$

$$\therefore \; f(x) = 2x^2 - 6x + 3$$

5.11 Definite Integrals

An indefinite integral:

$\int (x^2 + 2)\,dx$

A definite integral:

$\int_1^3 (x^2 + 3)\,dx$

Differentiating $f(x)$ gives us a new function from which we can determine the gradient at any point on the graph of $f(x)$. Similarly, integrating $f(x)$ results in a new function from which we can find the area under the graph. To do this we must define the vertical lines which bound the area. The x values are called the *limits* and, when included, result in a *definite* integral. The integration process is the same as before, but then follow the steps necessary to evaluate the area.

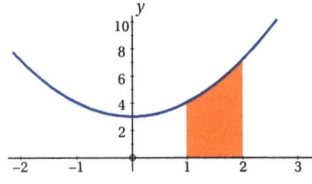

The diagram shows a shaded area bounded by the x-axis, the graph $y = x^2 + 3$, and the lines $x = 1$ and $x = 2$. The area is calculated by evaluating the definite integral $\int_1^2 x^2 + 3 \, dx$. Although you will have been shown how to do the calculation by hand, in the exam you can always find definite integrals on your GDC. In this case, area $= \int_1^2 x^2 + 3 \, dx = 5.33...$

The diagram shows the graph of $f(x) = (x-1)^2 \times a^x + b$ for $0 \le x \le 5$ and $a > 0$.

P is the y-intercept, Q is a local minimum and R is a local maximum.

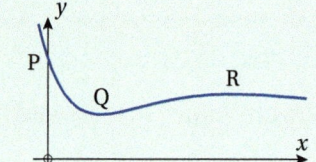

(a) Given that P $= (0, 2)$, determine the value of b.

(b) Given that $f(2) = 1.25$, show that $a = 0.5$.

(c) Find the coordinates of Q and R.

(d) For what values of x is f a decreasing function.

(e) Find the area under the curve between Q and R giving your answer to 2DP.

(a) $\quad 2 = (-1)^2 \times a^0 + b$ $\qquad 2 = 1 + b$ $\qquad b = 1$ (b) $\quad 1.25 = 1^2 \times a^2 + 1$ $\qquad 0.25 = a^2$ $\qquad a = 0.5 \,(\text{since } a > 0)$ (c) $\quad$ Q $= (1, 1)$, R $= (3.885, 1.563)$ (d) $\quad 0 \le x < 1,\; 3.885 < x \le 5$ (e) $\quad$ Area $= \int_1^{3.885} (x-1)^2 \times 0.5^x + 1 \, dx = 3.8556...$ $\qquad\qquad \approx \; 3.86$	*(a) It is tempting to say that $b = 2$ since this is the constant term in the function. But you must substitute the x-coordinate into the function.* *In (d), note the careful use of inequality signs, and also the need to refer to the domain.* *(e) Don't just put the answer – the examiner will want to see the integral written out before you use the GDC to calculate the area.*

Area below the axis: If the area is *below* the x-axis then the integral will have a negative value – but the area, of course, will still be expressed as a positive number. For example, $\int_0^1 x^3 - 2 \, dx = -1.75$ but the area under the curve is 1.75.

Area between a curve and the y-axis: In this case the integral must be written as $\int_a^b f(y)\,dy$ and this will probably involve rewriting the function.

Example: Find the area enclosed by the curve of $f(x) = \sqrt{x^2 + 2}$, the lines $y = 2$, $y = 3$ and the y-axis.

Solution: If $y = \sqrt{x^2 + 2}$ then $x = \sqrt{y^2 - 2}$ so the area is $\int_2^3 \sqrt{y^2 - 2}\,dy = 2.05$.

Volumes of revolution: The method of finding the area under a curve can be extended to calculate the volume generated when part of a curve is rotated through 360° around either the x-axis or the y-axis.

The diagram shows the shape generated when the part of the curve of $y = f(x)$ lying between $x = a$ and $x = b$ is rotated around the x-axis. Imagine a cross-section of the shape at a distance x from the origin; it will be a disc.

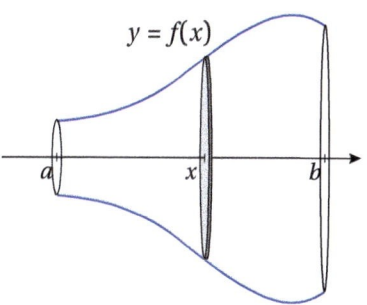

What is its volume?

- Its radius is $f(x)$
- So its cross-sectional area is $\pi\{f(x)\}^2$
- If its width is dx, its volume is $\pi\{f(x)\}^2\,dx$

dx is used for a very small distance in the x direction

The overall volume will be the sum of an infinite number of such discs, and hence is found by integration.

$$V = \int_a^b \pi\{f(x)\}^2\,dx$$

For clarity, this is usually written as $V = \int \pi y^2\,dx$

Areas and Volumes: Quick Practice

1. Find the area enclosed by the curve $y = 2xe^x$, the x-axis and the lines $x = -2$ and $x = 0$.

2. Find the area enclosed by the curve $y = x^2 - 1$, the y-axis and the lines $y = 1$ and $y = 2$.

3. Find the volume generated when the curve $y = \dfrac{3}{x + 2}$ is rotated through 360° about:

   (a) The x-axis between $x = 1$ and $x = 2$

   (b) The y-axis between $y = 1$ and $y = 1.5$

4. $f(x) = 4x^3 - 8x^2 - 3x + 9$

   (a) Find the roots of the equation $f(x) = 0$

   (b) Hence calculate the area enclosed by the curve of f and the x-axis.

Answers

1. 1.188
2. 1.578
3. (a) 2.356
   (b) 0.216
4. (a) $x = -1$, $x = 1.5$
   (b) 13.0

5.12 Kinematics (2)

We first looked at kinematics problems in the differentiation section on page 191. Now we can cope with a full range of likely questions by using integration as well.

When position, velocity and acceleration are defined as functions of time, we can move from one to the other using the fact that velocity is rate of change of position, and acceleration is rate of change of velocity.

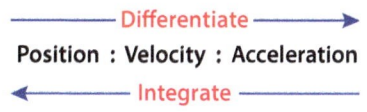

Look back to the problem involving the ball thrown up into the air, but let's suppose we were given the velocity function $v = 20 - 10t$, and not the height function. We integrate this to find position (height) and find that $h = 20t - 5t^2 + c$. Given that $h = 0$ when $t = 0$, we deduce that $c = 0$ and that $h = 20t - 5t^2$.

If we use a *definite* integral on v we will find the change in position over the given time interval, but we do need to be careful if there is a change of direction, as in fact there is at $t = 2$. This is why:

Change in position $t = 0$ to $t = 2$ is:

$$\int_0^2 20 - 10t \, dt = [20t - 5t^2]_0^2 = 20 - 0 = 20$$

Change in position $t = 2$ to $t = 3$ is:

$$\int_2^3 20 - 10t \, dt = [20t - 5t^2]_2^3 = 15 - 20 = -5$$

The negative change in position indicates we have reversed direction. So the integral from $t = 0$ to $t = 3$ will be $20 - 5 = 15$ which is its change in position compared to the start of the motion. If we want to know the total distance travelled we need to do the two separate integrals, then add the 20 upwards to the 5 downwards to get 25 m overall.

A neat way of doing this in one go using the GDC is to make use of the "absolute value" function. See if you can set this up on your GDC:

$$\int_0^3 |20 - 10t| \, dt$$

You should get 25, any negatives having been turned into positives. (This is in your formula book.)

Some questions give velocity as a function of displacement, not time. Using the chain rule, $\dfrac{dv}{dt} = \dfrac{dv}{ds} \times \dfrac{ds}{dt} = v\dfrac{dv}{ds}$. Thus the acceleration can be expressed as $v(s)\dfrac{dv}{ds}$.

Example: The acceleration in ms^{-2} of a particle moving in a straight line at time t seconds, $t \geq 0$, is given by the formula $a = 6t^2 + e^t$. The particle is at rest at $t = 0$.

    (a) Find an expression for v in terms of t.

    (b) What is the value of v when $t = 2.5\,s$?

Solution: (a) $v = \int 6t^2 + e^t \, dt = 2t^3 + e^t + c$

            When $t = 0$, $v = 0$ (since the particle is at rest at $t = 0$)

            $0 = 0 + 1 + c$ ∴ $c = -1$

            $v = 2t^3 + e^t - 1$

    (b) When $t = 2.5$, $v = 2 \times 2.5^3 + e^{2.5} - 1 = 42.4\,\text{ms}^{-1}$

5.13 Differential Equations

$\dfrac{dy}{dx}$ is called a *differential* and any equation containing one is a *differential equation*. We can solve an equation of the form $\dfrac{dy}{dx} = f(x)$ by integrating both sides wrt x (with respect to x) leading to the solution $y = F(x)$. For example, $\dfrac{dy}{dx} = 3x^2 + 1$ integrates to give the solution $y = x^3 + x + c$. This is called the *general solution*.

If we have been given a point on the solution curve, (1, 3) for example, we can substitute that to get c. This leads to the *particular solution* which in this case is $y = x^3 + x + 1$. There are two other types of first order differential equation to grapple with:

- $\dfrac{dy}{dx} = g(y)$. The problem here is that you cannot integrate the RHS wrt x because the letter is wrong. A little fiddle sorts this out: take $g(y)$ over to the LHS, and bring dx up to the RHS. For example:

$$\frac{dy}{dx} = e^y \;\Rightarrow\; \frac{dy}{e^y} = 1dx$$

$$\int \frac{1}{e^y} dy = \int 1dx \;\Rightarrow\; -e^{-y} = x + c$$

The variables may not be x and y, but it doesn't matter. The important thing is to make sure letters are not mixed up together when you integrate.

- $\dfrac{dy}{dx} = f(x)g(y)$. In this case, we still take the $g(y)$ down and bring the dx up, and then the $f(x)$ can team up with the dx. This will only work if the variables can be separated, and it may be necessary to do some algebra first. For example:

$$\frac{dy}{dx} = 2xy + 2x \;\Rightarrow\; \frac{dy}{dx} = 2x(y + 1)$$

$$\int \frac{1}{y + 1} dy = \int 2x dx$$

$$\ln(y + 1) = x^2 + c$$

If the variables cannot be separated we must use a numerical method, such as Euler's method on page 202.

In a case like this you may have to go further and make y the subject:

$$y + 1 = e^{x^2 + c}$$

$$y = A\,e^{x^2} - 1$$

You must be prepared to use **any** of the integral techniques you have learnt when solving differential equations.

Note how the +c becomes the multiplying constant A. This is because $e^{x^2 + c} = e^{x^2} \times e^c$ and e^c is itself a constant. Note too that it is only ever necessary to add a constant on one side of the equation.

Setting up a differential equation:

You may be lucky in that an exam question will give you the differential equation to solve – but it's also possible that you will have to set it up yourself. The key is to look for words such as "rate of change" or "rate of increase" which translate to the differential you will use. Here are some examples:

- "The rate at which a plant's height increases is proportional to the square root of its height" becomes $\dfrac{dh}{dt} = k\sqrt{h}$ (where k is the constant of proportionality).

- "A cup of boiling water cools at a rate of 0.006 times its temperature above room temperature, which is 20°C" becomes $\dfrac{d\theta}{dt} = -0.006(\theta - 20)$.

The rate of change is negative because the temperature (θ) is decreasing.

- "The acceleration is proportional to the reciprocal of the velocity" becomes $\dfrac{dv}{dt} = \dfrac{k}{v}$.

  (Use $a = v\dfrac{dv}{dx}$ if distance is involved).

Let's expand the last example into an exam-style question:

The acceleration of an object is inversely proportional to its velocity at any time. The initial velocity is $1\,ms^{-1}$ at which time the acceleration is $1.5\,ms^{-2}$.

(a) Find a differential equation to model this situation.

(b) Given that $v > 0$, and the initial velocity is $2\,ms^{-1}$, solve the differential equation, expressing v in terms of t.

(c) Calculate when the object reaches a velocity of $10\,ms^{-1}$.

Solution:

(a) $\dfrac{\mathrm{d}v}{\mathrm{d}t} = \dfrac{k}{v}$ and when $v = 1$, $\dfrac{\mathrm{d}v}{\mathrm{d}t} = 1.5$

So $1.5 = \dfrac{k}{1} \Rightarrow k = 1.5$

The differential equation is $\dfrac{\mathrm{d}v}{\mathrm{d}t} = \dfrac{1.5}{v}$

(b) $\int v \,\mathrm{d}v = \int 1.5 \,\mathrm{d}t \Rightarrow \dfrac{v^2}{2} = 1.5t + c$

When $t = 0$, $v = 1$ so $c = 0.5$

$\dfrac{v^2}{2} = 1.5t + 0.5$

$v^2 = 3t + 1$

$v = \sqrt{3t + 1}$

(c) $10 = \sqrt{3t + 1} \Rightarrow t = 33\,\text{s}$

A sample of radioactive material decays at a rate which is proportional to its mass, m. Given that 50 g of material decays to 48 g in 10 years, find the "half-life", ie the time taken for a given mass to halve.

$\dfrac{\mathrm{d}m}{\mathrm{d}t} = -km \quad \Rightarrow \int \dfrac{\mathrm{d}m}{m} = \int -k \,\mathrm{d}t$	*Once again the rate of change is negative so we must put a minus sign on the RHS of the differential equation.*
$\ln m = -kt + c$	
$m = e^{-kt+c}$	
$m = A e^{-kt}$	*This question is typical of those which lead to exponential growth or decay – make sure you try a few and become familiar with the method of solution.*
When $t = 0$, $m = 50$	
So $A = 50$, thus $m = 50\,e^{-kt}$	
When $t = 10$, $m = 48$	
So $48 = 50\,e^{-k \times 10} \Rightarrow k = 0.00408$	*We have both a constant of proportionality and a constant of integration, so we need two facts to find them; it only looks like one fact in the question but "hidden" is the initial mass.*
Solution of the differential equation is $m = 50\,e^{-0.00408t}$	
We require $m = 25$, so $25 = 50\,e^{-0.00408t} \Rightarrow t = 169.889$	
The half-life is 170 years.	

5.14 Slope Fields

Consider the differential equation $\dfrac{\mathrm{d}y}{\mathrm{d}x} = 2x$. We know that for every value of x, the solution curve has a gradient which is double the x-coordinate at any point. But, since the equation doesn't tell us anything about the y-coordinate of the point, it can be anything. For example, when $x = 1$ the gradient is 2, but there are an infinite number of possible points. This leads us to draw a *slope diagram* where we plot the gradient at a selection of points – without, of course, covering all the possibilities.

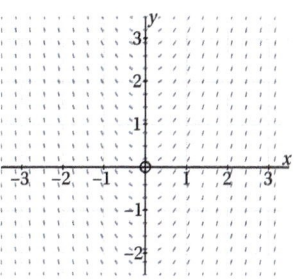

If I were to draw enough points that they practically join up, I would end up with a "family" of solution curves. Solving the equation gives us the general solution $y = x^2 + c$, each possible value of c giving us a different curve.

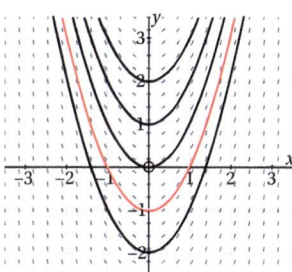

I've shown some of the family of curves in the next diagram, as well as the particular solution where $c = -1$ which is $y = x^2 - 1$.

If the differential equation is of the form $\dfrac{dy}{dx} = f(x, y)$ then at each point the slope depends on both coordinates.

Here, for example, is the slope field for $\dfrac{dy}{dx} = x - y$. The particular solution I have indicated in red is the line $y = x - 1$. We can verify that this is a solution of the equation by substitution.

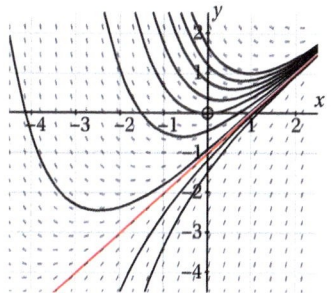

Note that no two curves in a family of curves can intersect. This is because, for any point in the plane, only one value of the slope is possible.

LHS: If $y = x - 1$ then $\dfrac{dy}{dx} = 1$

RHS: $x - y = x - (x - 1) = 1$

We can also glean other useful facts from the slope field. For example, all the solution curves have turning points where $\dfrac{dy}{dx} = 0$ and hence $y = x$. This line divides the graph into two regions: all points below $y = x$ have coordinates where $x > y$ and hence positive gradients. And all points above $y = x$ have negative gradients.

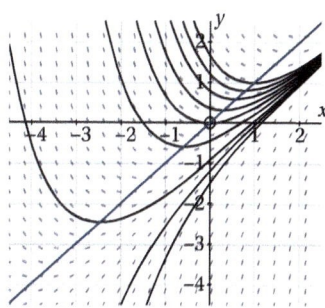

Use of the second derivative: Differentiation of $\dfrac{dy}{dx} = x - y$ gives $\dfrac{d^2y}{dx^2} = 1 - \dfrac{dy}{dx} = 1 - x + y$.

Thus the second derivative = 0 at points along the line $y = x - 1$; this is the red line in the first diagram. It therefore follows that the second derivative is positive for all points above the line, leading to solution curves which are concave upwards. And all curves below the line are concave downwards.

199

See page 204 for a description of the approximation method (Euler) for solving coupled differential equations.

Exact solution of coupled differential equations: Coupled differential equations occur when two quantities are changing, and their rate of changes are dependent on each other. In this context, we can find exact solutions for coupled differential equations of the form:

$$\frac{dx}{dt} = ax + by, \quad \frac{dy}{dt} = cx + dy.$$

The solutions are found as follows:

1. Write the coupled equations in the matrix form $\begin{pmatrix} \dot{x} \\ \dot{y} \end{pmatrix} = \begin{pmatrix} a & b \\ c & d \end{pmatrix} \begin{pmatrix} x \\ y \end{pmatrix}$

2. Find the eigenvalues and eigenvectors of $\begin{pmatrix} a & b \\ c & d \end{pmatrix}$

3. The solution is $\begin{pmatrix} x \\ y \end{pmatrix} = A e^{\lambda_1 t} \boldsymbol{p}_1 + B e^{\lambda_2 t} \boldsymbol{p}_2$ where the eigenvalues are λ_1 and λ_2 with corresponding eigenvectors $\boldsymbol{p}_1$ and $\boldsymbol{p}_2$. This solution is in your formula book, but note carefully the format which is used.

Example: Find the solution of the following system of differential equations:

$\dot{x} = x + 2y$, $\dot{y} = 4x + 3y$ given that when $t = 0$, $x = 4$ and $y = 5$.

Solution: In matrix form, $\begin{pmatrix} \dot{x} \\ \dot{y} \end{pmatrix} = \begin{pmatrix} 1 & 2 \\ 4 & 3 \end{pmatrix} \begin{pmatrix} x \\ y \end{pmatrix}$. The eigenvalues are found from the characteristic equation $(1 - \lambda)(3 - \lambda) - 8 = 0 \Rightarrow \lambda = -1$ or 5. Corresponding eigenvectors are $\begin{pmatrix} 1 \\ -1 \end{pmatrix}$ and $\begin{pmatrix} 1 \\ 2 \end{pmatrix}$. Thus the general solution is:

$$\begin{pmatrix} x \\ y \end{pmatrix} = A e^{-t} \begin{pmatrix} 1 \\ -1 \end{pmatrix} + B e^{5t} \begin{pmatrix} 1 \\ 2 \end{pmatrix}$$

It would be a useful exercise for you to check that these solutions for x and y do in fact satisfy the original coupled equations.

Now we put in the initial conditions, and get: $4 = A + B$, $5 = -A + 2B$, and solving these simultaneous equations gives us $A = 1$, $B = 3$. And, after a lot of huffing and puffing, we find that $\begin{pmatrix} x \\ y \end{pmatrix} = e^{-t} \begin{pmatrix} 1 \\ -1 \end{pmatrix} + 3 e^{5t} \begin{pmatrix} 1 \\ 2 \end{pmatrix}$.

Note in the worked example above that for very large values of t the solutions tend to $x = 3e^{5t}$ and $y = 6e^{5t}$. In other words, in the long term, the ratio of x to y will be 1:2.

Phase portraits: A diagram which plots the values of x and y over time is called a *phase portrait*; different initial values lead to different *trajectories* for x and y. The following facts will help you both to interpret a given phase portrait, and also to sketch one if required.

- A point where both $\dot{x}$ and $\dot{y}$ are zero is known as an *equilibrium point*. If nearby points tend to move away from the equilibrium point it is *unstable*; if they move towards it, it is *stable*. (Imagine a hemispherical bowl: the bottom of the bowl is a stable equilibrium point because a bead placed there will not move, and a bead placed a little way up the bowl will roll down to the point. But turn the bowl upside down, and the top point becomes unstable – any bead placed near to it will roll away from it).

- A *saddle point* is best illustrated with a diagram. In this diagram the origin is a saddle point, so called because the trajectories around it form the shape of a saddle.

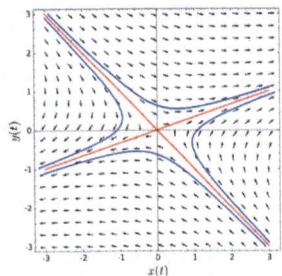

- Eigenvalues. The phase portrait illustrates the solutions of coupled differential equations, and the eigenvalues which were found during the process determine aspects of the trajectories. All the trajectories tend towards the two lines through the origin which are parallel to the eigenvectors (again, see diagram). Also:

  - If the two eigenvalues are positive, the origin is an unstable equilibrium point.

  - If the two eigenvalues are negative, the origin is a stable equilibrium point.

- If the one eigenvalue is positive and the other is negative, the origin is a saddle point. Look at the diagram (which is turning out to be rather useful!) and you'll see that all the trajectories move towards one of the lines and away from the other.
- Eigenvalues are calculated from a quadratic equation, so it is possible they are complex conjugates of the form $a \pm bi$. In which case:

    If $a = 0$ the trajectories form circles or ellipses around the origin.

    If $a > 0$ the trajectories spiral *away* from the origin (illustrated).

    If $a < 0$ the trajectories spiral *into* the origin.

Note that they move towards the line parallel to the eigenvector derived from the *positive* eigenvalue.

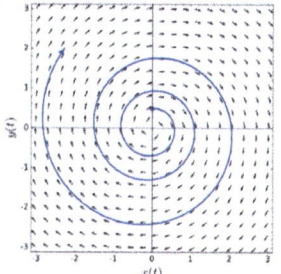

(a) Find the eigenvalues for the system of equations

$$\dot{x} = 3x + 4y$$

$$\dot{y} = -2x - y$$

The phase portrait for the equations in part (a) is a spiral around the origin.

(b) State whether the spiral is towards or away from the origin.

(c) Explain why $\dfrac{dy}{dx} = 0$ at points along the line $y = -2x$

(d) Find $\dfrac{dy}{dx}$ at the points (1, 0) and (0, 1)

(e) Sketch the phase portrait. Include the line $y = -2x$

(a) Eigenvalues found by $\begin{vmatrix} 3 - \lambda & 4 \\ -2 & -1 - \lambda \end{vmatrix} = 0$

$(3 - \lambda)(-1 - \lambda) + 8 = 0$

$\lambda^2 - 2\lambda + 5 = 0$

$\lambda = 1 \pm 2i$

(b) Away from the origin

(c) $\dfrac{dy}{dx} = \dfrac{\dot{y}}{\dot{x}} = \dfrac{-2x - y}{3x + 4y}$

Thus $\dfrac{dy}{dx} = 0$ when $-2x - y = 0 \Rightarrow y = -2x$

(d) At (1, 0) $\dfrac{dy}{dx} = -\dfrac{2}{3}$

At (0, 1) $\dfrac{dy}{dx} = -\dfrac{1}{4}$

(e)

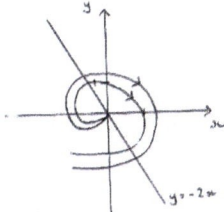

Ensure your GDC is setup to display results in a + bi form, otherwise the quadratic will yield no solutions.

In part (c) the chain rule gives us

$$\frac{dy}{dx} = \frac{dy}{dt} \times \frac{dt}{dx} = \dot{y} \times \frac{1}{\dot{x}}$$

In part (e) we can also use gradients to decide whether the spiral is in or out.

5.15 Euler's Method

Solving $\frac{dy}{dx} = f(x,y)$:   In the same way that the trapezoidal rule is used to find the area under the graph of a function which we can't integrate, so Euler's method is used to solve a differential equation numerically: this is useful when the variables cannot be separated.

Consider the differential equation in the previous section, $\frac{dy}{dx} = x - y$. Let's suppose we take the solution curve which passes through the point (–1, 3), and we want to find y when $x = 1$. We **do** know the gradient at (–1, 3): it's $x - y = -4$. Euler's method takes a small step in the x direction, let's say $\delta x = 0.5$, and assumes the curve carries on as a straight line to the next point. We then take the gradient at the new point, and carry on with a series of straight lines until we get to the required x value. The diagram below shows the method graphically, compared to the actual curve on the slope diagram.

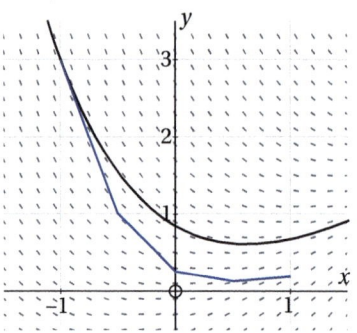

Noting that the change in y, δy, is calculated as $\delta x \times \frac{dy}{dx}$, we can draw up a table to do the calculations:

x	y	$\frac{dy}{dx} = x - y$	$\delta y = 0.5\frac{dy}{dx}$
–1	3	–4	–2
–0.5	1	–1.5	–0.75
0	0.25	–0.25	–0.125
0.5	0.125	0.375	0.1875
1	0.3125		

This is not particularly accurate because we have taken such a large step length. Decrease it to 0.1 and we end up at (1, 0.6088) which is much closer to the actual solution which is (1, 0.6765).

In a chemical reaction between two substances p and q it is thought that the rate of change of mass (g) of p compared to q can be modelled by the differential equation $\dfrac{\mathrm{d}q}{\mathrm{d}p} = 2q - 4p$. The diagram shows the slope field for the differential equation, and a solution curve with initial value $(0, k)$ where $k = 1 - \dfrac{1}{e}$.

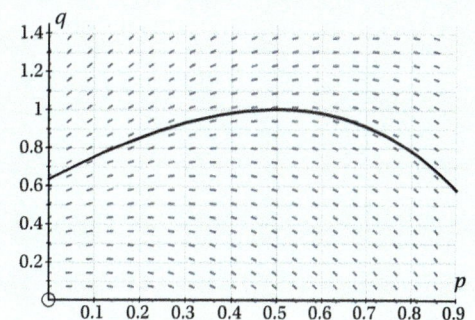

(a) Find the local maximum value of q on the solution curve given that it occurs when the mass of p is 0.5 g.

(b) (i) Show that $q = 2p + 1$ is a solution of the differential equation.

(ii) Find $\dfrac{\mathrm{d}^2 q}{\mathrm{d}p^2}$ in terms of p and q.

(iii) Hence explain why the mass of q for the given solution curve must eventually drop to zero.

(c) (i) Calculate the value of $1 - \dfrac{1}{e}$ to 4 decimal places.

(ii) Use Euler's method with steps in p of 0.1 to estimate the value of q when $p = 0.3$ given that initially the mass of $p = 0$ and the mass of $q = 1 - \dfrac{1}{e}$.

(a) $q' = 2q - 4p$

$0 = 2q - 4 \times 0.5 \Rightarrow q = 1$.

(b) (i) $q' = 2$, but $q' = 2q - 4p = 2(2p + 1) - 4p = 2$

(ii) $q'' = 2q' - 4 = 2(2q - 4p) - 4 = 4q - 8p - 4$

(iii) $q'' = 0$ when $4q - 8p - 4 = 0 \Rightarrow q = 2p + 1$

Points below the line will be concave downwards, and since all points on the solution curve are below $q = 2p + 1$, there will be no more turning points. Thus the curve will intercept the p-axis.

(c) (i) 0.6321

(ii)

p	q	$\dfrac{\mathrm{d}q}{\mathrm{d}p} = 2q - 4p$	$\delta q = 0.1\dfrac{\mathrm{d}q}{\mathrm{d}p}$
0	0.6321	1.2642	0.12642
0.1	0.7585	1.1171	0.11171
0.2	0.8703	0.94050	0.094050
0.3	0.9643		

When $p = 0.3$, $q = 0.9643$

The important thing to remember throughout questions such as these is that the first and second derivative have the significance they always have. Thus they can be used to find turning points, the nature of the curve at a point and so on.

When doing the calculations for Euler's method, always work to more figures than you need for the eventual answer.

I have shown the step-by-step solution for completeness. In an exam you would not be expected to draw up a table like this, but set up your GDC with the relevant iterative formula, and then just show the answer. However, you should show the formula you are using to gain method marks.

Coupled differential equations: We saw in the previous section how to find the exact solution of a particular type of coupled differential equation. Using Euler's method, we can find approximate solutions for the more general system where:

$$\frac{dx}{dt} = f_1(x, y, t) \text{ and } \frac{dy}{dt} = f_2(x, y, t)$$

The step-by-step solutions use the following formulae, which are also in your formula book:

$$t_{n+1} = t_n + h$$

$$x_{n+1} = x_n + hf_1(x_n, y_n, t_n)$$

$$y_{n+1} = y_n + hf_2(x_n, y_n, t_n)$$

Coupled differential equations which model predator-prey relationships offer good examples: the numbers of predators and prey clearly depend on each other. Let's consider an example where x and y represent the number of prey in thousands and the number of predators in hundreds. The differential equations are $\frac{dx}{dt} = 4x - 2xy$ and $\frac{dy}{dt} = xy - 3y$, with initial values $x = 2$ and $y = 2$. Time is in years.

Before looking at the Euler method for solution, here are some other possible questions you may be asked:

- Find the two equilibrium positions, and what they each represent.

Equilibrium is when rate of change is zero. Putting each of the differentials equal to zero gives us two possibilities: $x = 0$, $y = 0$ and $x = 3$, $y = 2$. The first represents extinction of both species; the second represents a long-term population of 3000 prey and 200 predators which will never change.

- Is the predator population initially increasing or decreasing?

Substituting $x = 2$ and $y = 2$ into $xy - 3y$ gives $\frac{dy}{dt} = -2$, so initially the number of predators is decreasing at 200 per year.

- If the predators became extinct, how fast does the model predict that the prey population would increase?

With $y = 0$, $\frac{dx}{dt} = 4x$. They would quadruple every year (assuming discrete rather than continuous growth).

Now let's use Euler's method to predict the populations after 1 year with step lengths of $t = 0.2$.

Try setting this up on your GDC. You should use a sequence function for successive values of t.

t	x_n	y_n	$x_n + 0.02(4x_n - 2x_ny_n)$	$y_n + 0.02(x_ny_n - 3y_n)$
0	2	2	2	1.96
0.02	2	1.96	2.0032	1.9208
0.04	2.0032	1.9208	2.009546	2.009546
0.06	2.009546	2.009546	2.01899	1.845216
0.08	2.01899	1.845216	2.031491	1.809013
1	2.031491	1.809013		

Thus, after 1 year, the prey has increased by about 31, the predators have decreased by about 19.

If the prey numbers continue to increase, we would expect the predator numbers to start increasing as well – and that is what happens as this graph of the numbers for the first 4 years shows.

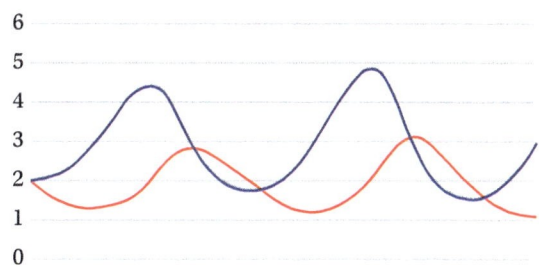

Second order differential equations: Only second order equations of the form $\dfrac{d^2y}{dx^2} = f\left(x, \dfrac{dx}{dt}, t\right)$ are considered in this course. If we substitute $y = \dfrac{dx}{dt}$ then, because $\dfrac{dy}{dt} = \dfrac{d^2x}{dt^2}$, we can replace the second order equation with the system of equations $\dfrac{dy}{dt} = f(x, y, t)$ and $\dfrac{dx}{dt} = y$. These can then be solved using Euler's method for coupled equations.

For example, suppose $\dfrac{d^2x}{dt^2} + 2\dfrac{dx}{dt} + 2x = 6\cos 3t$.

- Let $y = \dfrac{dx}{dt}$

- Then $\dfrac{d^2x}{dt^2} = \dfrac{dy}{dt} = -2y - 2x + 6\cos 3t$

...and the Euler method then gives:

- $t_{n+1} = t_n + h$

- $x_{n+1} = x_n + hy$

- $y_{n+1} = y_n + h(-2y_n - 2x_n + 6\cos 3t_n)$

Try setting up a table where $h = 0.1$ and initial values are $t = 0$, $x = 2$, and $y = 1.5$. You should find that when $t = 0.3$, $x = 2.385$.

Differential Equations: Practice Exercise

1. Find the particular solution of the following differential equations. In each case make y the subject.

   (a) $\dfrac{dy}{dx} = \dfrac{2x+1}{3y^2}$, $y = 2$ when $x = 1$

   (b) $\dfrac{dy}{dx} = 4x(y+1)$, $y = 10$ when $x = 0$

   (c) $\dfrac{dy}{dt} + y = 3$, $y = 2$ when $t = 0$

2. (a) Show that for a particle moving with velocity v at time t along a straight line, its acceleration can be expressed as $v\dfrac{dv}{dx}$, where x is its displacement from the origin.

   (b) At time t the acceleration of the particle is given by $a = \cos x$. Given that $v = 3$ when $t = \dfrac{\pi}{2}$, find an expression for v in terms of x.

Answers

1. (a) $y = (x^2 + x + 6)^{\frac{1}{3}}$
   (b) $y = 11\,e^{2x^2} - 1$
   (c) $y = 3 - e^{-t}$
2. (a) $a = \dfrac{dv}{dt} = \dfrac{dv}{dx} \times \dfrac{dx}{dt} = v\dfrac{dv}{dx}$
   (b) $v = \sqrt{2\sin x + 7}$
3. (a) $\begin{pmatrix} x \\ y \end{pmatrix} = Ae^{0.1t}\begin{pmatrix} 1 \\ 1 \end{pmatrix} + Be^{0.4t}\begin{pmatrix} 1 \\ 2 \end{pmatrix}$
   (b) $A = 0.6$, $B = 1$.
   (c) $y = Ae^{0.3t}$
4. (a) 10.862...
   (c)(i) 50 (ii) 9.0635...
   (iii) 50
   (d)(i) 19.8% (ii) Shorter step length

3. (a) Find the general solution of the following system of differential equations:

$$\dot{x} = 0.2x + 0.1y$$

$$\dot{y} = 0.2x + 0.3y$$

   (b) Find the values of the constants A and B given that when $t = 0$, $x = 1.6$ and $y = 2.6$

   (c) If in the long term the value of x reduces to 0, find the new equation for y in terms of t.

4. The acceleration of a parachutist in free fall (before he opens his parachute) is given by $\dfrac{dv}{dt} = 10 - 0.2v$, where v is his velocity in ms^{-1} t seconds after leaving the plane. Assume that when $t = 0$, $v = 0$.

   (a) Use Euler's method with step length of $t = 0.2$ to find a value for v when $t = 1$.

   (b) Show by substituting that the general solution of the differential equation is $v = 50 - A\,e^{-\frac{1}{5}t}$

   (c) (i) Find A

      (ii) Hence find his actual velocity after 1 second

      (iii) State the value of his long-term velocity

   (d) (i) Find the percentage error in the approximate value

      (ii) How could Euler's method yield a more accurate result?

Calculus: Long Answer Questions

See www.peakib.com

The answers to the following Paper 2 style questions are given; full working can be found on the Peak Study Resources website.

1. A circular oil slick on the surface of the sea had a radius of 45 m when first observed, and its radius was increasing at 0.8 m/min. At a time t minutes later, its radius is r metres.

    If left untreated, its radius will increase at a rate which is inversely proportional to the square of its radius.

    (a) (i) Explain why the differential equation $\dfrac{\mathrm{d}r}{\mathrm{d}t} = \dfrac{k}{r^2}$ represents the situation.

    (ii) Find the value of k.

    (iii) Hence solve the differential equation to find r in terms of t.

    (iv) Calculate the radius of the oil slick after 1 hour, if left untreated.

    It is planned to treat the oil slick with detergent, and the rate of increase of radius is then expected to be proportional to $\dfrac{1}{r^2(2+t)}$ with the same initial conditions.

    (b) (i) Write down a differential equation for the new situation involving a constant of proportionality m.

    (ii) Find the value of m.

    (iii) Calculate the new value of the predicted radius after 1 hour.

    _____

    Answers:

    (a) (ii) $k = 1620$

    (iii) $r = \sqrt[3]{4860t + 91125}$

    (iv) 72.6 m

    (b) (i) $\dfrac{\mathrm{d}r}{\mathrm{d}t} = \dfrac{m}{r^2(2+t)}$

    (ii) $m = 3240$

    (iii) 49.9 m

2. The diagram shows part of the graph of $f(x) = -x^3 + 3x + 2$.

A is the y-intercept, B is a local maximum and C is the x-intercept.

The diagram also shows the straight line (AB).

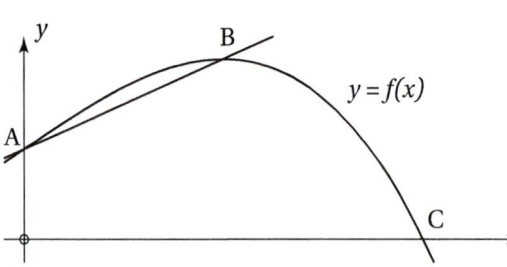

(a) Write down the coordinates of A.

(b) Write down $f'(x)$.

(c) Use your answer to part (b) to show that the x-coordinate of B is 1, and find the y-coordinate of B.

(d) Find the equation of (AB) in the form $ax + by + c = 0$

(e) (i)   Find the area enclosed by the graph of f, the x-axis and the vertical lines through A and B.

     (ii) Find the area of the trapezium formed by the line segment [AB], the x-axis and the vertical lines through A and B.

     (iii) Hence find the area enclosed between the line and the curve.

(f) Find the coordinates of C.

(g) (i)   Copy and complete the following table of values of $f(x)$

x	0	0.5	1	1.5	2
$f(x)$		3.375			

     (ii) Use the trapezoidal rule with 4 strips to find an approximation to the area under the graph of f from A to C.

Answers:

(a) A = (0, 2)

(b) $f'(x) = -3x^2 + 3$

(c) B = (1, 4)

(d) $2x - y + 2 = 0$

(e) (i)   3.25

     (ii) 3

     (iii) 0.25

(f) C = (2, 0)

(g) (i)   Missing values are: 2, 4, 3.125, 0

     (ii) 5.75

3. A cyclist is cycling up a hill. His speed ($\dot{x}$ ms⁻¹) t seconds after he begins the climb can be modelled by the equation $\dot{x} = 4 + \dfrac{2}{0.8t + 1}$, where x is the distance from the start of the climb.

    (a) Write down his initial speed.

    (b) Find an expression for his acceleration.

    (c) Explain why the cyclist's speed is always decreasing.

    (d) Find an expression for the distance travelled by the cyclist after t seconds.

    (e) The hill is 200m long. Find the cyclist's speed at the top of the hill.

    The cyclist descends the hill using the same road. His speed is given by the equation $\dot{x} = 3 + e^{-0.5t}(5t - 1)$.

    (f) Find his maximum velocity.

    (g) Find the time it takes to reach the foot of the hill.

    _____

Answers:

(a) $6\,\text{ms}^{-1}$

(b) $\ddot{x} = \dfrac{-1.6}{(0.8t + 1)^2}$

(c) Acceleration is always negative, or velocity is a decreasing function

(d) $x = 4t + 2.5\ln(0.8t + 1)$

(e) $4.05\,\text{ms}^{-1}$

(f) $6.29\,\text{ms}^{-1}$

(g) $60.7\,\text{s}$

4. The diagram below shows the slope field for the differential equation $\frac{\mathrm{d}P}{\mathrm{d}t} = P - 2t$.

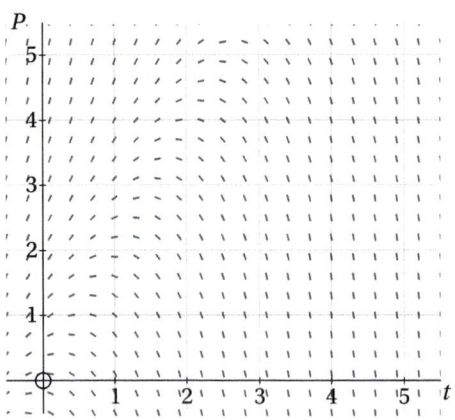

(a) Verify that $P = 2t + 2$ is a solution to this differential equation.

(b) Draw a set of axes with $0 \le t \le 5$ and $0 \le P \le 5$ and plot $P = 2t + 2$

Let P represent the population density of a bacteria (for $0 \le P \le 5$) in a patient t days after the bacteria enters their system. If initially $P(0) < a$ the population will reduce to 0.

(c) State the minimum value of a.

(d) By considering the gradient of all points below the line $P = 2t + 2$ explain why if $P(0) < a$ the trajectory can never cross the line $P = 2t + 2$

(e) On your axes sketch the trajectory of the population density if at t = 0,

(i) $P = 1.5$ (ii) $P = 2.5$

A patient will experience symptoms if the density of bacteria is ever greater than 2.7.

(f) Use Euler's method with a step length of 0.2 to see when a patient will first experience symptoms if $P(0) = 1.5$

Answers:

(b)

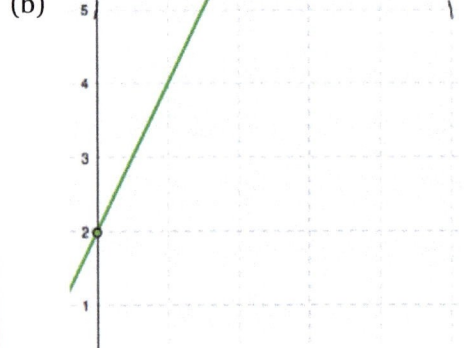

(c) 2

(d) For all points such that
$P < 2t + 2$, $\frac{\mathrm{d}P}{\mathrm{d}t} < 2t + 2 - 2t = 2$

Hence the gradient is always less than 2 but to approach the line the gradient has to be greater than 2.

(e)

(f) Approximately 1 day

5. Two products, X and Y, are competing in the same market After t weeks the number of product X being sold per week is x, and the number of product Y being sold is y. The growth in the number of each product sold can be modelled by the following coupled differential equations:

$$\dot{x} = 0.4x - 0.1y$$

$$\dot{y} = -0.6x + 0.5y$$

for $x, y \geq 0$.

(a) Find the eigenvalues for the system.

(b) Hence, find the general solution to the system of equations.

(c) Draw the phase diagram for this system, for $x, y \geq 0$.

Initially X has sales of 100 per week.

(d) Write down a condition on the number of sales of Y in that week if Y is to become the only product on the market.

(e) In the case where X initially has sales of 100 per week and Y has sales of 300 per week,

    (i) find the time at which the sales of X reduce to 0

    (ii) the sales of Y at this time.

Answers:

(a) 0.2 and 0.7

(b) $\begin{pmatrix} x \\ y \end{pmatrix} = A\,e^{0.2t}\begin{pmatrix} 1 \\ 2 \end{pmatrix} + B\,e^{0.7t}\begin{pmatrix} 1 \\ -3 \end{pmatrix}$

(c)

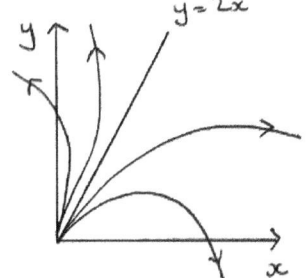

(d) $y > 200$

(e) (i) 3.58 years

    (ii) 1226

Chapter 6: MAXIMISING YOUR MARKS

Remember that the examiner is on your side – they want to give you marks! Make it easy for them to follow your thinking, even if you are not quite sure what you are doing or if you are getting wrong answers. You cannot lose marks for doing things wrong. LEARN THIS CHECKLIST.

Before you start a question:

- Read it carefully so you know what it is about.
- Highlight important words.

Answering a question:

- Check any calculations you do, preferably using a different method or order of operation.
- Show your working – there are often marks for method as well as for the right answer. And, in a longer question, a wrong answer at the start may mean lots more wrong answers – but the examiner will probably give you marks for correct methods, and will check your working against your original answer.
- Make sure you have answered *exactly* what the question asked. For example, have you been asked to calculate the new value of an investment or the amount of interest earned.
- In longer questions, don't worry if you can't work out the answer to a part. Carry on with the rest, using their answer (if one is given) or even making up a reasonable answer.

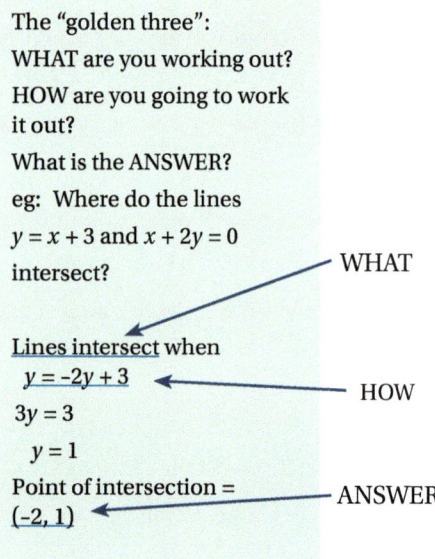

The "golden three":
WHAT are you working out?
HOW are you going to work it out?
What is the ANSWER?
eg: Where do the lines
$y = x + 3$ and $x + 2y = 0$
intersect? — WHAT

Lines intersect when
$y = -2y + 3$ — HOW
$3y = 3$
$y = 1$
Point of intersection =
$(-2, 1)$ — ANSWER

- Don't spend too long on any question or part of a question – you may lose the opportunity to answer easier questions later on. You can always come back and fill in gaps. Work to one mark per minute!
- Use words to explain what you are doing, especially in a longer question.
- The algebra can be tough – keep going!
- Check the units in questions – are they mixed?

Diagrams:

- Do not assume facts from diagrams, especially if they are marked NOT TO SCALE. For example, it may *look* like a right angle but does the question *tell* you that it is? Two lines may *look* parallel but they aren't unless you are *told* they are.
- And do draw your own diagrams – not necessarily to hand in as part of the question, but to help you sort out what's going on.

Key words in questions ("command terms"):

STATE: put the answer down without working (should be an easy one)!

WRITE DOWN: minimal working required.

SHOW: show enough working to get to the given answer.

EVALUATE: give a value to, work out.

SKETCH A GRAPH: draw its shape and show key points (eg: where it cuts the axes)

PLOT A GRAPH: work out points and draw the graph accurately

EXACT VALUE: not a rounded decimal eg: 2π, not 6.28...

SHOW $x = 3$ is the solution of $2x + 1 = 7$.

$$2 \times 3 + 1 = 7$$

(We have not had to *solve* the equation)

When you have answered the question:

- Check you have answered every part of the question.
- Check you have answered exactly what was asked.
- Check you have answered to the correct accuracy (normally 3SF)
- Check that what you have written is clear, and that your answer is not mixed up in the working somewhere.

DO THESE CHECKS – you will probably pick up a few marks.

Chapter 7: PRACTICE QUESTIONS

The questions which follow are not designed to cover every aspect of the syllabus, nor are they exam style questions. Their purpose is to give you some practice in the *basics*: if you cannot, for example, carry out a straightforward differentiation, then you will get questions which depend on accurate differentiation wrong, even if you know exactly how to do the question. So you need to answer all these questions as part of your revision. If you get an answer wrong, find out why; then come back to it later, and see if you can get it right next time.

NUMBER AND ALGEBRA

1. Calculate $\dfrac{4.523}{6.81 \times 2.47}$ giving your answer to 3SF.

2. A triangle has a base of 12.5 cm and a height of 7.8 cm, both measurements to 1DP. What is the maximum possible area?

3. What is the percentage error if the fraction $\dfrac{22}{7}$ is used for π?

4. If a computer carries out 5.6×10^9 calculations per second, how long does 1 calculation take? Give your answer in standard form.

5. Convert 58 km/h into m/s.

6. Find the 25th term and the sum of the first 54 terms of the sequence which begins: 3, 8, 13, 18...

7. An arithmetic sequence has first term 7 and common difference 3.5. How many terms are required for the sum of the sequence to be 25 830?

8. What is the 12th term and the sum to 18 terms of the sequence which begins 3, 12, 48, 192?

9. A geometric series has a first term 400, ten terms and a sum of 1295.67. What is the common ratio?

10. What is the sum of the series $\displaystyle\sum_{r=1}^{r=6}(3 \times 2^r - r^2)$

11. How much will an investment of \$6300 be worth (to the nearest dollar) after accumulating compound interest for 12 years at a rate of 3% per annum? If 1.5% interest is paid every 6 months, how much will the investment be worth after 12 years?

12. Solve the equation $-2x^4 + 3x^2 = -4$ for $x > 0$. Answers to 3DP.

13. Find the solution of the simultaneous equations $2y = 3x + 5$ and $6x + 3y = 11$.

14. Solve the system of equations:

$$\begin{cases} 3x - 6y + 2z = 14 \\ x + y + z = 0 \\ -12x + 1z = 3 \end{cases}$$

15. Write $2 + 3\log_{10}x$ as a single logarithm.

16. If $s = 3 + 10e^{0.4t}$, find t in the form $a \ln b$ when $s = 15$.

17. Find the eigenvalues and the eigenvectors for the matrix $M = \begin{pmatrix} 0.4 & 0.25 \\ 0.6 & 0.75 \end{pmatrix}$. By expressing M as a product of three matrices, find an expression for M^n, and the long term stable value of M^n.

18. Convert to cis form: $1 + i$, $2 - 3i$, $-6i$, $(2 + i)^2$

19. Convert to $a + bi$ form: $2 \operatorname{cis} 60°$, $3(\cos\frac{5\pi}{3} + i \sin\frac{5\pi}{3})$, $4e^{i\pi}$

20. Find the values of z such that $z^2 = 5 - 12i$.

21. $f(x) = 3 \sin\left(2x + \frac{\pi}{6}\right)$ and $g(x) = 4 \sin\left(2x + \frac{3\pi}{4}\right)$.

    Express $f(x) + g(x)$ in the form $a \sin(2x + b)$. State the amplitude and period of $f(x) + g(x)$.

FUNCTIONS

1. What is the range of the function $f(x) = x^2 + 2$, $x \geq 1$.

2. Sketch the graph of $f(x) = \sqrt{2 - x}$, and hence write down the domain and range of the function. Add the graph of $f^{-1}(x)$ to your sketch, and state its range.

3. Write down the gradients, x and y intercepts of the following straight line functions: $y = 2x - 5$; $3x - 2y = 8$; $x + y = 10$

4. Work out the equation of the line perpendicular to $y = 3x - 4$, which goes through the point $(1, 1)$.

5. Find the line of symmetry and the vertex of the graph of $y = x^2 - 2x + 6$.

6. For the graph of $y = 2(x - 3)(x + 1)$, write down the coordinates of the x intercepts, the y intercept and the vertex.

7. Some radioactive material decays such that its mass m after t years is given by the formula $m = 12 \times 2^{-0.008t}$. What is its mass at $t = 0$ and after 100 years?

8. The graph with equation $y = 3^{2x} + k$ passes through the point $(1, 6)$. Find the value of k and find x when $y = -2$.

9. Solve the following equations: $x + 3 = \frac{2}{x}$; $x^3 = 2^x$.

10. Find the horizontal and vertical asymptotes on the graph of $y = \frac{2 + x}{3 - x}$.

11. Sketch the graph of $y = 0.5x^3 - \frac{2}{x}$, $-2 \leq x \leq 2$, $x \neq 0$. Mark any axis intercepts, and name the equation of the vertical asymptote.

12. For the graph of $f(x) = \frac{e^{-x}}{(x + 1)^2}$, identify any horizontal and vertical asymptotes. Find the turning point, and the solutions of the equation $f(x) = 7$.

13. Solve $3.1^x = 10^{x-1}$ giving your answer to 4DP.

14. Write down the minimum and maximum values, and the period, of the graph of $y = 2\sin 12x + 3$.

15. The kinetic energy of a body, measured in Joules, is directly proportional to the square of its speed in ms⁻¹. If a body travelling at $20\,\text{ms}^{-1}$ has kinetic energy of $1000\,\text{J}$, calculate the kinetic energy of a body travelling at $25\,\text{ms}^{-1}$.

16. The population $P(t)$ of fish in a large lake t years after measurements begin is given by the logistic model $P(t) = \dfrac{10000}{1 + 19\,e^{-kt}}$. Find the population when $t = 0$. Given that there are 1500 fish when $t = 1$, find k. What is the largest fish population the lake can accommodate?

17. It is thought that the set of data points (1.1, 3.0), (2.3, 4.2), (3.0, 4.8) can be modelled by a function of the form $y = ax^k$. Use logarithms base 10 to linearise the model, draw a graph and hence find the values of a and k.

18. For the function $f(x) = 0.5x^3 - 1.1x + 2.5$, find the x- and y-intercepts; the gradient of the line AB where A and B are the vertices; and the equation of the graph when $f(x)$ is transformed by a stretch ×2 parallel to the y-axis followed by a reflection in the x-axis.

19. The height of the tide is measured against the harbour wall in Eastport on a single day. Above the seabed, the maximum height is 8.1 m and the minimum height is 1.9 m. The maximum height occurs at 0200, and the minimum height at 0800. The suggested model for the height h, in metres, at time t hours after midnight is of the form $h = a + b\cos(k(t - d))°$. Find values for a, b, k, and d.

GEOMETRY AND TRIGONOMETRY

1. For the points A(2,3) and B(4,7), find: the midpoint; the distance AB; the gradient of AB; the equation of AB.

2. Repeat question 1 for the points A(−1,4) and B(−3,7).

3. Write down which pairs of lines are parallel and which are perpendicular:
   (a) $y = 2x + 5$; (b) $y + 2x = 1$; (c) $x = 4$; (d) $2y = x$; (e) $y - 2x + 3 = 0$; (f) $y = 3$.

4. Triangle PQR has a right angle at Q. PQ = 12 cm, R = 23°. Calculate PR.

5. Triangle ABC has a right angle at A. AC = 2.4 cm, AB = 1.8 cm. Find C.

6. Solve the following triangles (the triangle in each case is ABC):

   BC = 6 cm, C = 87°, A = 45°. Find AB.

   AB = 6 cm, A = 87°, AC = 5.4 cm. Find BC.

   AB = 6 cm, BC = 5.4 cm, CA = 3.5 cm. Find B.

   AB = BC = 5.2 cm. B = 34°. Find AC.

   AC = 6 cm, C = 32°, A = 90°. Find AB.

   BC = 6 cm, B = 62°, C = 71°. Find AC.

7. Find the area of the first and second triangles in question 6.

8. A line AB of length 60 m is marked out on a river bank and the position of a tree C on the opposite bank is surveyed. It is found that CAB = 58° and CBA = 38°. Find CA and the width of the river.

9. Calculate the volume and surface area of a sealed cylinder with radius 3.5 cm and height 8.3 cm.

10. Calculate the volume and surface area of a sphere with diameter 8 m.

11. Calculate the radius of a sphere with volume 128 cm³.

12. Calculate the curved surface area of a cone with radius 5 cm and height 8 cm.

13. A square based pyramid has AB = 10 cm and AE = 12 cm. Q is the midpoint of AB. Calculate the following:

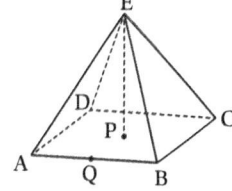

    The lengths AC, AP, PE, QE.

    The angle between AE and the base.

    The angle between QE and the base.

    The angle between EB and EC.

    The surface area of the pyramid (including the base).

14. A sector of a circle of radius 4.5 cm has an area of 35.34 cm². Find the angle of the sector to the nearest degree, and also the perimeter of the sector.

15. A = (2, –2), B = (6, 0), C = (0, 4). Find the equations of the perpendicular bisectors of AB and BC, and also the point where the two bisectors meet.

16. The arc length of a sector of a circle of radius 2.25 cm has length $\frac{27}{16}\pi$. Find the area of the sector, giving your answer in an exact form.

17. Show that the equation $\sin^2\theta = 1 - \sin^2\theta$ is equivalent to $\tan^2\theta = 1$. Solve the equation for $-\pi < \theta < \pi$.

18. For the transformation with matrix $\begin{pmatrix} 4 & 1 \\ 3 & 2 \end{pmatrix}$, find the image of the point (1, –3), and the point which transforms to (10, 10).

19. Write down the matrix which performs a rotation around the origin of +45°, writing each element in surd form. Also write down the matrix which performs an enlargement of $\sqrt{2}$ centre the origin, and hence the matrix which is equivalent to a +45° rotation around the origin followed by an enlargement of $\sqrt{2}$ centre the origin.

20. A model plane follows a straight line path at constant velocity. It passes through the points (3, 0, 4) and (6, –6, 10) (distances in m). Find a vector equation for its path; its height after 3 seconds; its speed; the point where it intersects the path of another plane which is $r = \begin{pmatrix} 7 \\ 2 \\ -3 \end{pmatrix} + s\begin{pmatrix} 0 \\ -2 \\ 3 \end{pmatrix}$. Do the planes collide?

21. $a = \begin{pmatrix} 1 \\ -1 \\ 4 \end{pmatrix}$ and $b = \begin{pmatrix} 6 \\ -2 \\ k \end{pmatrix}$: find the angle between them if $k = 2$; the value of k if a and b are perpendicular. If $k = 1$ find the vector product $a \times b$ and the component of a acting perpendicular to the direction of b, in the plane of a and b.

22. Which algorithms would you use to find: shortest route round weighted graph; minimum spanning tree; least weight Hamiltonian cycle; upper bound for the travelling salesman problem.

STATISTICS AND PROBABILITY

The amount spent (in €) by the first 50 people going into a shop is shown in the table below:

15.60	5.95	31.22	3.02	6.60	24.70	15.45	32.50	12.45	4.43
12.65	10.09	52.86	12.88	2.53	31.79	9.86	25.79	18.28	32.05
14.87	24.65	15.70	8.65	4.42	17.20	8.53	0.45	0.95	4.44
7.45	5.82	45.20	2.70	10.04	15.70	32.20	12.43	36.75	32.50
16.87	3.78	0.56	33.67	9.67	25.50	33.06	7.56	2.63	45.80

Questions 1 to 9 refer to this table.

1. Is this data discrete or continuous?

2. Draw up a grouped frequency table (with first group €0.01 – €10.00). You should have 6 groups.

3. Which is the modal group?

4. Enter the mid-values of each group and the frequencies onto your GDC. Calculate estimates of the mean and the standard deviation. (Why "estimates"?)

5. Draw a bar chart to represent the data.

6. Complete a cumulative frequency table for the data, and hence draw a cumulative frequency graph.

7. From the cumulative frequency graph, write down the median, the lower quartile, the upper quartile and calculate the interquartile range.

8. Draw a box plot for the data. Is the maximum value an outlier?

9. What was the least amount that the people in the top ten percentiles spent?

10. The mean of the numbers 1, 7, 8, 10, 11 and $k - 2$ is k. What extra number must be added to increase the mean to $k + 1$?

11. Use the data in the following table to calculate the correlation coefficient and the equation of the regression line of y on x.

x	1	4	4	6	8	10	11	12
y	30	28	36	30	39	35	40	44

12. Two dice are thrown. What is P(at least one shows a number greater than 1)?

13. I have 6 red socks and 4 green socks in a drawer. I take 2 out at random. Draw a tree diagram to show the possible outcomes and find P(the two socks do not match).

14. A and B are two events. P(A) = 0.2, P(B) = 0.5 and P(A ∪ B) = 0.55. Use a Venn diagram to find: P(A ∩ B); P(A′ ∩ B); P(A|B); P(B′|A).

15. Given that $P(A) = \frac{2}{3}$, $P(B|A) = \frac{2}{5}$ and $P(B|A') = \frac{1}{4}$, find P(B′) and P(A′ ∪ B′).

16. Two dice are rolled. Find the probability that they show different numbers given that the total is 8.

17. Given that P(A ∪ B) = 0.7, P(A) = 0.6 and that A and B are independent events, find P(B).

18. The probability distribution for a discrete random variable X is as follows:

x	1	2	3	4	5
$P(X = x)$	0.3	0.35	k	$2k$	0.05

Find the value of k and the expected mean.

19. For $X \sim B(12, 0.2)$, find $P(X = 3)$, $P(X \leq 2)$, $P(X > 4)$. What is the mean of X?

20. If $X \sim N(100, 5^2)$, find $P(X < 112)$, $P(X < 91)$, $P(95 < X < 101)$.

21. An apple grower finds that the weight of apples from his orchard are normally distributed with mean 85 g and standard deviation 15 g. He classes the heaviest 12% of his crop as "large". What is the weight of the least heavy large apple?

22. Find the Spearman rank correlation coefficient for the following data which shows the number of years a group of 11 people have been in their current job against their ages.

Years in job	2.1	3.5	0.6	1.0	2.4	5.8	1.7	2.0	6.3	4.9	3.5
Age	25	38	26	30	31	38	22	24	30	27	25

23. A particular machine in a factory occasionally fails. The average number of failures X in a 7 day week can be modelled by a Poisson distribution where $X \sim P_0(0.24)$. Find the probability that: there are no failures in a week; there are exactly 2 failures in 2 weeks; there are at least 2 failures in 28 days; there is at least 1 failure in each of 3 successive weeks.

24. X is an independent random variable such that $X \sim B(12, 0.25)$. Find $E(X)$, $var(X)$, $E(2X + 1)$, $var(3X - 4)$.

25. $Y \sim P_0(4.5)$. Write down the distribution of the sample mean $\bar{Y}$ for a sample size of 20, and the probability that a sample of size 20 has a mean less than 4.

26. The marks for 8 students taking a Maths exam and a French exam were as follows:

Maths	52	25	86	33	55	57	54	46
French	40	48	65	57	41	39	63	34

Calculate the product moment correlation coefficient and Spearman's rank correlation coefficient for the data.

27. A hypothesis test at the 5% level is carried out on the binomial distribution $B(10, 0.35)$. $H_0: p = 0.35$, $H_1: p \neq 0.35$. Number of successes is 7. Find the p value and state the conclusion.

28. This set of data comes from a normal distribution: 2.8, 3.5, 4.1, 3.7, 1.2, 3.9, 2.5. Test at the 5% level whether it could have come from a normal distribution $N(3.3, 0.3^2)$.

29. Test at the 10% level whether the set of observed frequencies {7, 13, 10, 15, 4, 2} (which relate to X values {0, 1, 2 ...}) could have come from a Poisson distribution with mean 2.2.

30. Test at the 5% level and at the 1% level whether these two samples could come from normal distributions with the same mean:

Sample 1: Mean = 12.8, SD = 1.45, sample size = 60

Sample 2: Mean = 13.4, SD = 1.5, sample size = 80

CALCULUS

1. Differentiate the following functions: (a) $x + \frac{6}{x}$ (b) xe^{-x} (c) $2\ln(\cos x)$
   (d) $\frac{x^2 - 2}{x}$ (e) $\frac{3x^3}{(x+1)}$ (f) $\sin x \tan x$ *(answer in terms of $\sin x$)* (g) $\sqrt{x^3 - 2}$

2. Find the equation of the tangent and the equation of the normal to $y = 3x^2 + 3$ at the point where $x = 1$.

3. Find the two points on the graph of $y = x - \frac{2}{x}$ where the gradient $= 3$.

4. By finding the values of $f(x)$ and $f'(x)$ at the points where $x = 2$ and $x = 3$ on the graph of $y = \frac{1}{2}x^2 - 2x + 1$, sketch the shape of the graph between the two points.

5. Find the turning point and the point of inflexion on the graph of $y = x\,e^x$. *(Exact answers)*

6. Use your calculator to find all the turning points on the graph of $f(x) = x^2 \times 2^x$. State which are maxima and which are minima. Give answers to 3 significant figures where appropriate.

7. The perimeter of a rectangle is a cm. If its width is x cm, find its length in terms of a and x. Prove that the area is a maximum when all sides are the same length.

8. A circular oil slick is increasing in radius at the rate of 2 m/min. Find, in terms of π, the rate at which the area of the slick is increasing when its radius is 30 m.

9. Find the area enclosed by the curve $y = (x^2 - 3)\ln x$ and the x-axis.

10. Integrate these functions: (a) $\int \sin 3x\,dx$ (b) $\int \frac{4x}{x^2 - 1}\,dx$ (c) $\int 2e^{2x} + 3\ dx$
    (d) $\int x\sqrt{x^2 - 3}\ dx$

11. Find $f(x)$ if $f'(x) = \frac{3}{x^2}$ and $y = 3$ when $x = 1$.

12. Use the trapezoidal rule with 4 strips to find an estimate for $\int_2^4 \frac{e^x}{x^2}\,dx$. Find the percentage error of this estimate.

13. Find the real number $k > 1$ for which $\int_1^k \left(1 + \frac{1}{x^2}\right)dx = \frac{3}{2}$.

14. A particle moves in a straight line. At time t secs its acceleration is given by $a = 3t - 1$. When $t = 0$, the velocity of the particle is 2 ms^{-1} and it is 3 m from the origin. Find expressions for v and s in terms of t. Show that the particle is always moving away from the origin.

15. Find the particular solution of the differential equation $x^2\frac{dy}{dx} = y + 3$, $y = -2$ when $x = 1$.

16. Use Euler's method to find y when $x = 1$, given that $\frac{dy}{dx} = x^2 + y^2$, and that $y = 0$ when $x = 0.5$. Use a step length of $x = 0.1$.

17. Find the eigenvalues for the system of equations: $\dot{x} = 2x + y$, $\dot{y} = -x + 2y$. Determine if the solution path is spiralling away from or towards the origin.

18. By substituting y for $\dot{x}$, use Euler's method to solve the differential equation $\ddot{x} + 2\dot{x} - 3x = 6t$, given that $y = 2$ when $x = 1$. Use a step length of $t = 0.1$, and find approximate values of x and y when $t = 0.5$.

Answers to Practice Questions

NUMBER AND ALGEBRA

1. 0.269
2. $49.25875 \, cm^2$
3. 0.0402%
4. 1.79×10^{-10}
5. $16.11 \, m/s$
6. 123, 7317
7. 120
8. $12\,582\,912$, 6.87×10^{10}
9. 0.7
10. 285
11. \$8982, \$9006
12. 0.922, 1.533
13. $x = \frac{1}{3}$, $y = 3$
14. $x = \frac{2}{3}$, $y = -1\frac{2}{3}$, $z = 1$

15. $\log_{10} 100 x^3$
16. $2.5 \ln 1.2$
17. $0.15, 1; \begin{pmatrix} -1 \\ 1 \end{pmatrix}, \begin{pmatrix} \frac{5}{12} \\ 1 \end{pmatrix};$

$$\begin{pmatrix} \frac{1}{4}\left(3\left(\frac{1}{5}\right)^n + 1\right) & -\frac{1}{4}\left(\left(\frac{1}{5}\right)^n - 1\right) \\ -\frac{3}{4}\left(\left(\frac{1}{5}\right)^n - 1\right) & \frac{1}{4}\left(\left(\frac{1}{5}\right)^n + \frac{3}{4}\right) \end{pmatrix};$$

$$\begin{pmatrix} \frac{1}{4} & \frac{1}{4} \\ \frac{3}{4} & \frac{3}{4} \end{pmatrix}$$

18. $\sqrt{2} \operatorname{cis} 45°$, $\sqrt{13} \operatorname{cis} 56°$, $6 \operatorname{cis}(-90°)$, $5 \operatorname{cis} 53.1°$.
19. $1 + i\sqrt{3}, \frac{3}{2} - \frac{3\sqrt{3}}{2} i, -4$
20. $\pm(3 - 2i)$
21. $4.335 \sin(2x + 1.624)$; 4.335; π

FUNCTIONS

1. $f(x) \geq 3$
2. $x \leq 2, f(x) \geq 0$; $x \leq 2$

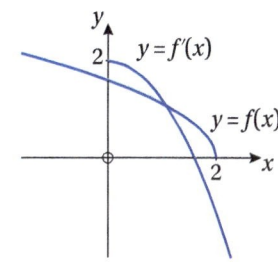

3. $2, 2.5, -5; 1.5, \frac{8}{3}, -4; -1, 10, 10$
4. $x + 3y = 4$
5. $x = 1$, $(1,5)$
6. $(3,0), (-1,0); (0,-6); (1,-8)$
7. $12, 6.89$
8. $-3, 0$
9. 0.562 and $-3.56, 1.373$
10. $y = -1, x = 3$

11.

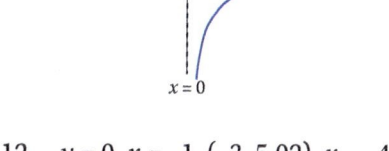

12. $y = 0$, $x = -1$, $(-3, 5.02)$, $x = -4.38$ or -0.512 or -2.06
13. 1.9660
14. $-1, 5, 30°$
15. $1562.5 \, J$
16. $500; -1.210; 10000$
17. $\log_{10} y = k \log_{10} x + \log_{10} a$; $a = 2.8$, $k = 0.5$
18. $-2.132; -0.733; y = -x^3 + 2.2x + 5$
19. $5, 3.1, 20, 2.$

GEOMETRY AND TRIGONOMETRY

1. $(3,5)$, 4.47, 2, $y = 2x - 1$.

2. $(-2,5.5)$, 3.61, -1.5, $2y + 3x = 5$

3. $\parallel$ a,e. $\perp$ c,f; b,d.

4. 30.7

5. $36.9°$

6. 8.47, 7.86, $35.3°$, 3.04, 3.75, 7.24

7. 18.9, 16.2

8. 37.1, 31.5

9. $319.4 \, \text{cm}^3$, $259.5 \, \text{cm}^2$

10. $268.1 \, \text{m}^3$, $201.1 \, \text{m}^2$

11. $3.13 \, \text{cm}$

12. $148.2 \, \text{cm}^2$

13. $14.1 \, \text{cm}$, $7.07 \, \text{cm}$, $9.70 \, \text{cm}$, $10.9 \, \text{cm}$; $53.9°$; $62.7°$; $49.2°$; $318 \, \text{cm}^2$

14. $200°$, $24.71 \, \text{cm}$

15. $2x + y = 12$, $3x - 2y = -4$; $\left(\frac{20}{7}, \frac{44}{7}\right)$

16. $\frac{243}{128}\pi$

17. $-\frac{\pi}{4}, \frac{\pi}{4}$

18. $(1, -3)$, $(2, 2)$

19. $\begin{pmatrix} \frac{1}{\sqrt{2}} & -\frac{1}{\sqrt{2}} \\ \frac{1}{\sqrt{2}} & \frac{1}{\sqrt{2}} \end{pmatrix}, \begin{pmatrix} \sqrt{2} & 0 \\ 0 & \sqrt{2} \end{pmatrix}, \begin{pmatrix} 1 & -1 \\ 1 & 1 \end{pmatrix}$

20. $r = \begin{pmatrix} 3 \\ 0 \\ 4 \end{pmatrix} + t\begin{pmatrix} 1 \\ -2 \\ 2 \end{pmatrix}$, $10 \, \text{m}$, $3 \, \text{ms}^{-1}$, $(7, -8, 12)$; No ($t = 2$, $s = 5$).

21. $22.5°$, -2. $\begin{pmatrix} -7 \\ -25 \\ -8 \end{pmatrix}$, 4.24

22. Chinese postman; travelling salesman; Prim or Kruskal; nearest neighbour.

STATISTICS AND PROBABILITY

1. Discrete.

2.

0.01 – 10.00	10.01 – 20.00	20.01 – 30.00	30.01 – 40.00	40.01 – 50.00	50.01 – 60.00
20	14	5	8	2	1

3. 0.01 – 10.00

4. 17.2, 13.31; not using original data.

5.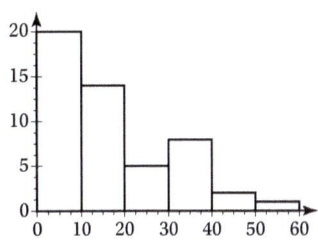

6.

€	≤10	≤20	≤30	≤40	≤50	≤60
c.f.	20	34	39	47	49	50

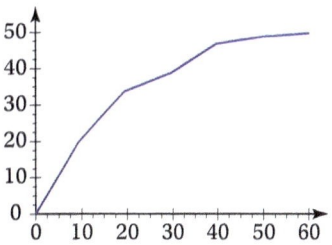

7. $Q_1 = 6$, $Q_2 = 14$, $Q_3 = 28$, IQR = 22

   No. $Q_3 + 1.5 \times$ IQR = 61

8.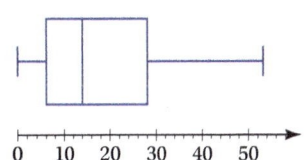

9. €38

10. 14

11. 0.796, $y = 1.15x + 27.2$

12. 35/36

13. P(no match) = 8/15; see diagram:

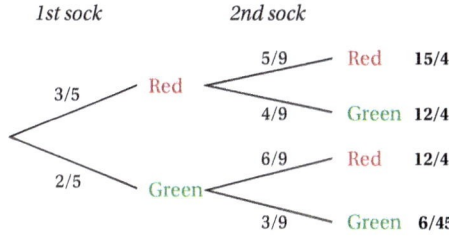

14. 0.15, 0.35, 0.3, 0.25

15. $\dfrac{13}{20}, \dfrac{11}{15}$

16. 0.8

17. 0.25

18. 0.1, 2.35.

19. 0.236, 0.558, 0.0726; 2.4

20. 0.992, 0.0359, 0.421

21. 102.6 g

22. 0.486

23. 0.787, 0.0713, 0.250, 0.00971

24. 3, 2.25, 7, 7.75

25. $\overline{Y} \sim N\left(4.5, \dfrac{4.5}{20}\right)$, 0.146

26. 0.304, 0.167

27. 0.0261; Accept H_0

28. p value = 0.077, yes it could.

29. p value = 0.204 > 0.1. Values could have come from the distribution.

30. Significant at the 1% level but not at the 5% level.

CALCULUS

1. (a) $1 - \dfrac{6}{x^2}$

   (b) $e^{-x}(1-x)$

   (c) $-2\tan x$

   (d) $1 + \dfrac{2}{x^2}$

   (e) $\dfrac{3x^2(2x+3)}{(x+1)^2}$

   (f) $\dfrac{\sin x}{1-\sin^2 x} + \sin x$

   (g) $\dfrac{3x^2}{2\sqrt{x^3-2}}$

2. $y = 6x,\ x + 6y = 37$

3. $(1,-1),\ (-1,1)$

4.

5. $(-1, -e^{-1}),\ (-2, -2e^{-2})$

6. $(-2.89, 1.13)$ max, $(0,0)$ min

7. $\frac{1}{2}a - x$

8. 120π

9. 0.173

10. (a) $-\frac{1}{3}\cos 3x + c$

    (b) $2\ln(x^2-1) + c$

    (c) $e^2 x + 3x + c$

    (d) $\frac{1}{3}(x^2-3)^{\frac{3}{2}} + c$

11. $y = -\dfrac{3}{x} + 6$

12. $557.0365;\ 4.86\%$

13. 2

14. $v = 1.5t^2 - t + 2,$
    $s = 0.5t^3 - 0.5t^2 + 2t + 3;\ v \neq 0$ for any value of t (discriminant < 0). So v is always positive, and particle is moving away from the origin.

15. $y = e^{1-\frac{1}{x}} - 3$ *(other arrangements are possible).*

16. 0.2597

17. $1 \pm 2i$; spiralling away.

18. $x = 2.026,\ y = 2.639$